Energy Resources

Energy Resources

From Science to Society

FIRST EDITION

WESLEY REISSER
Adjunct Professor
George Washington University

COLIN REISSER

New York Oxford
OXFORD UNIVERSITY PRESS

Published in the United States of America by Oxford University Press
198 Madison Avenue, New York, NY 10016, United States of America.

Library of Congress Cataloging-in-Publication Data

Names: Reisser, Wesley J., author. | Reisser, Colin, author.
Title: Energy resources : from science to society / Wesley Reisser, adjunct professor, George Washington University, Colin Reisser.
Description: New York : Oxford University Press, [2019] | Includes index.
Identifiers: LCCN 2017058679 | ISBN 9780190200497 (pbk.)
Subjects: LCSH: Power resources—Popular works.
Classification: LCC TJ163.2 .R438 2019 | DDC 333.7—dc23 LC record available at https://lccn.loc.gov/2017058679

Printing number: 9 8 7 6 5 4 3 2 1
Printed by Sheridan Books, Inc. United States of America

Contents

Preface and Acknowledgments

The global energy system and its future are central to understanding many key challenges faced by society today. From climate change to rising prices for commodities to instability in the Middle East, energy resources touch our lives in many ways. Indeed, energy touches everything we do every day; from flipping the electrical switch to turn on a light to powering the tractors and trucks that produce and ship our food, modern life is reliant on vast quantities of energy to underpin everything we do. Yet, most of this energy remains hidden from the majority of the population. Most people are blissfully unaware of where the electricity they use to turn on the lights comes from, where the gasoline they pump into the car comes from, or how the fresh fruit on the grocery store gets to the shelf. We take cheap, plentiful, and reliable energy for granted, but a vast global system with tremendous impacts on all of us underpins it.

This book is intended to help fill a gap in our energy understanding. Students of engineering, geology, and physics study the underlying science and technology of energy, and some students of geopolitics (although far too few) study the global impacts of our energy choices. What is lacking is a broader literacy of all sides of the energy system, from a simple understanding of where resources come from to how we harvest and consume them and the subsequent impact on our politics, environment, economy, and society. This book aims to build that foundation in a way that is accessible to a broader audience. We present each resource and give all their due attention because it is impossible to understand the global energy system we have, let alone where it is going, if we ignore the fossil fuels that built the modern world and continue to mostly fuel it today.

After exploring each energy resource, we turn to cross-cutting energy issues we must consider as we work toward a cleaner energy future. Widespread energy literacy is essential to overcoming the challenges we face as our current energy system drives geopolitical strife, pollution, and indeed the very habitability of the planet we live on.

This project would not have been possible without the support and input of many people. It has been a true pleasure to work together as brothers, both geographers, on this joint project. Our parents, Kurt and Susan Reisser, provided essential edits and an initial audience to review the manuscript. Growing up with Kurt, a petroleum geologist as well as an avid hiker and lover of nature, helped us gain a keen sense of the energy system at a young age. And Susan, a writer and lover of literature, helped us learn to write in a manner that we hope you will find engaging, clear, and fun. Ken Regelson, editor of EnergyShouldBe.org, gave us an excellent outside scrub of the book from its early stages and provided many of the graphics found in the text. Ken's advice and knowledge helped us fill the gaps in our own knowledge on the fast-changing world of clean energy. Much of the structure of the course this book builds on, taught by Wesley Reisser at the George Washington University, came from professor emeritus Robert J. Weimer of the Colorado School of Mines. His material was invaluable to embarking on this project. The support of faculty at the George Washington University Department of Geography, especially Marie Price, Lisa Benton-Short, Elizabeth Chacko, Nikolay Shiklomanov, Michael Mann, and Dima Streletskiy, is greatly appreciated. Similarly, the advice and support of faculty at the UCLA Department of Geography, especially John Agnew, Thomas Gillespie, and Larry Smith, is acknowledged.

We also thank the many reviewers, both anonymous and those listed below, who reviewed various manuscripts over the past several years:

Jennifer Baka, Penn State University
Dorothy Boorse, Gordon College
Mary Brake, Eastern Michigan University
Tugrul Daim, Portland State University
Floyd Hayes, Pacific Union College
Delia Heck, Ferrum College
Gal Hochman, Rutgers University
Ryan Holifield, University of Wisconsin–Milwaukee
Brandon Hoover, Messiah College
Jordan P. Howell, Rowan University
Phil Kelly, Emporia State University
Ned Knight, Linfield College
Pankaj Lal, Montclair State University
Stephen MacAvoy, American University
Daniel Marien, University of Central Florida
Robert McCallister, University of Wisconsin–Rock County
Michael McElfresh, Santa Clara University
Mark Meo, University of Oklahoma
Enrique Lanz Oca, Hunter College (CUNY), BMCC (CUNY), and
      Vassar College
Rajesh Sharma, Arkansas State University
Benjamin Sovacool, Vermont Law School
Boyka Stefanova, University of Texas–San Antonio
Hongtao Yi, Ohio State University

It has been a true pleasure to work with Oxford University Press on this project. Our editor, Dan Kaveney, has steered us through the process. Christine Mahon and Megan Carlson have also been invaluable in helping us navigate through our first textbook. We also thank our production team led by Theresa Stockton, along with Holly Haydash, and Patricia Berube.

Disclaimer: The views expressed throughout this work are solely those of the authors and do not reflect the views of the U.S. Government.

Image 1.1 Mihai Maxim / Shutterstock

CHAPTER 1

What Is Energy?

Energy makes modern civilization possible. Its generation and supply have become increasingly complex as demand increases along with the world's population, making the development of secure, plentiful, and sustainable energy one of the biggest challenges for humanity in our era. At its most basic, energy is the ability to do work. Energy can be broken down into kinetic energy and potential energy, both of which will be investigated further in relation to the global economy and society. Kinetic energy allows for work, whether through propelling transportation modes, powering machines, heating and cooling spaces, or countless other processes. Potential energy is a measure of the ability to produce kinetic energy, and humanity has harnessed a variety of resources to create and store this potential energy, the basic underpinning of modern energy supply.

Energy and the resources involved in its production sit at the core of our modern way of life. Dramatic advances in the means to produce and consume energy over the past several hundred years have compressed the planet we live on by dramatically accelerating our ability to move people, goods, and ideas across the earth. Energy technologies have enabled human beings to live in ever more concentrated populations with greater comforts than ever before (Images 1.2 and 1.3). Energy technologies have also enabled billions of people to escape from hard physical labor by replacing human muscle with machine power.

Image 1.2 Energy-intensive machines, such as the introduction of refrigeration to preserve and store food, have vastly improved quality of life.
Everett Collection / Shutterstock

Image 1.3 Modern transportation systems use vast amounts of energy to move people and goods quickly over long distances.
Peter Stuckings / Shutterstock

These advances have come with major costs, however. In the world of energy production and consumption, there is no perfect solution. Our environment, society, and politics are dramatically affected by our energy systems. From conflicts over oil in the Middle East, deaths caused in the mining of energy resources such as coal, and severe environmental damage, energy resource production and consumption propel further instabilities and, in some cases, calamities. One of the greatest challenges that coming generations of human beings will confront, manmade climate change, has been driven primarily through the burning of fossil fuels (Image 1.4), and no solution is possible without a substantial retooling of our energy system.

Because of energy's centrality in our modern way of life, solving these global challenges entails major changes for how we produce and consume resources. We must investigate the production and consumption of energy resources at a variety of scales, from the local to the national to the global, to understand the links within and potential means to improve the energy system. We must look at solutions that are both appropriate and realistic because renewable energy technologies have not yet reached the level of development necessary to completely replace the fossil fuels on which our current way of life depends. To meet these myriad challenges, it is necessary to understand the science and the policy impacts of our choices, rather than dealing with these issues in isolation. This text is intended to be a handbook that brings together the scientific underpinnings of each energy resource with its greater social, economic, and environmental context. This multidisciplinary approach investigates the impacts of our energy choices while promoting an understanding of how each resource is produced and consumed, including how each resource fits into the global and national consumption patterns of energy. Finally, we will investigate cross-cutting energy resource challenges that extend beyond the realm of one resource, including transportation, pollution, climate change, and the geopolitics of energy. By looking at energy from science to society, we can better understand and hopefully address the challenges of our global energy system. Over the twenty-first century, addressing our production and consumption patterns for energy will be

Image 1.4 The modern energy system creates many drawbacks, including air and water pollution.

Panupong Ounhaphattana / Shutterstock

essential to coping with the climate crisis, growing populations, and an increasing appetite for the energy-intense lifestyle that people living in highly developed countries already enjoy.

Overview of Energy Basics

For our purposes in this text, we will consider **energy** the capacity/ability to do work. This means the ability to move something against its resistance or resting state. **Potential energy** measures the amount of stored potential to do work. Energy should not be confused with **power**, which is the rate of doing work. Power concerns the conversion of energy from one state to another or across space. **Kinetic energy** is the use of potential energy to produce motion or change (Image 1.5).

The most important energy concept lies in the **law of conservation of energy**. The law states that the quantity of energy remains constant and cannot be created or destroyed within a closed system. This means that there is a finite amount of energy present in our universe that simply changes form or location. Thus, if we start with one hundred units, we end with one hundred in some state or another. "Consuming" energy is about changing the state of an energy source to another state while harnessing the change for our use. The earliest human use of such a resource was the burning of biological materials, especially wood, in fires to produce heat. This change of state from a biological solid into heat energy is just one example of the application of this law in our use of energy.

Image 1.5 The boulder on the left has potential energy because of its elevation. On the right, the potential energy is converted to kinetic energy by gravity.

Olivier / Wikimedia Commons / Flickr / CC-BY-SA-3.0

In our current global energy system, we tend to use chemical energy to produce **heat energy** or kinetic energy. **Chemical energy** is the stored energy in the chemical bonds of compounds, such as those found in fossil fuels and biomass. When burned, chemical energy releases heat energy, which can be used in an engine to also produce kinetic energy and, through that, **electrical energy**. More modern energy storage methods such as batteries also take advantage of chemical energy, albeit in different ways. Other energy resources such as hydropower or wind power can be used to harness kinetic energy and convert it directly into electrical energy. No matter which form we look at, we are principally looking at energy *conversion*, not *consumption*, although in the vernacular we tend to refer to energy consumption throughout the text.

Our energy system can be broken down into three basic pieces: energy production, energy transmission, and energy consumption. We must also cope with energy storage during different phases of this process. The production end consists of collecting and harnessing energy resources for human use. The transmission segment involves moving the resource, in either its natural or its refined state, from the place of production to the place of consumption. This global movement of energy resources and electricity creates many of the challenges we face in the current energy system. Finally, the consumption end refers to the end use of different energy resources to perform a variety of key functions, most notably the production of electricity; the heating and cooling of homes, businesses, and factories; and the propelling of various modes of transportation.

Measuring Energy

Before investigating specific energy sources, it is essential to understand how we measure and convert these differing resources. Throughout this book, we will rely primarily on the prevailing standard unit for measuring a particular resource, most often metric units (with English Imperial units noted after). The standard basic units of measure, from which all others derive under the *Systeme Internationale d'Unites* (SI system), are the meter, kilogram, and second. From these, we derive the other basic measures. However, for certain energy resources, such as oil, the prevailing standard measure is not metric or tied to the SI system, and we will rely on the industry standard (barrels in the case of oil) for measurement.

The most basic energy unit in our energy system is the **joule** (J). A joule is the amount of energy expended to apply the force of 1 newton over the length of 1 meter. For our purposes, a joule can best be thought of as the energy required to produce 1 watt for 1 second. The **watt** (W) is the basic unit of power that we will use most frequently in this text. A watt measures the rate at which energy is converted from one form to another. One watt is a tiny amount of energy, and therefore we usually refer to it in far larger numbers. We do this primarily through the addition of a prefix to the term, as shown in Table 1.1.

Table 1.1 Unit Prefixes

Prefix	Symbol	Multiplies by
Kilo	k	One thousand
Mega	M	One million
Giga	G	One billion
Tera	T	One trillion
Peta	P	One quadrillion

For instance, we would refer to kilowatts (kW) or megawatts (MW) rather than watts when discussing electrical capacities to keep numbers more manageable. When we discuss global energy consumption, we prefer to use terawatts (TW) because of the enormous amount of energy consumed.

The amount of energy consumed over time is measured slightly differently. The kilowatt-hour (kWh) measures this consumption, with 1 kWh equaling the amount of energy converted in one hour at the rate of 1 kW. Thus, a 10 kW machine running for thirty minutes would convert 5 kWh of energy. When converted to joules, 1 kWh consists of 3.6 MJ (since there are 3,600 seconds in an hour). We tend to think of the kilowatt and kilowatt-hour in terms of electricity, since that is where they are most often referred to in popular discourse. However, these units can be used to measure any form of energy conversion, not just to or from electrical energy.

In the United States, the **British thermal unit** is often encountered, rather than joule- and watt-based quantities. One British thermal unit is the heat energy needed to raise 1 pound of water by 1 degree Fahrenheit, equal to 1,055 J. The quad or quadrillion British thermal unit is commonly used in the United States to measure energy quantities on the national scale. For the purposes of this book, we will avoid this unit of measure because of its currently limited use within the United States and a few other formerly British jurisdictions.

In global energy discussions, oil and coal equivalents are also discussed on occasion as a means to compare various energy resources. One **ton of oil equivalent** measures the amount of energy derived from burning 1,000 kg of oil, approximately 41.88 GJ. Prior to the use of ton of oil equivalents, it was more common to refer to 1 **ton of coal equivalent**, which is the heat released from burning 1 metric ton of coal, with 28 GJ being the international standard, although varying grades of coal burn at rates far below to slightly above this range.

For oil, some statistics use the ton of oil equivalent figure; however, most measures of oil production and consumption use the measurement of the **barrel**, or blue barrel. This measurement refers to the early days of oil production, when oil was stored in barrels (Image 1.6), which eventually became standardized at 42 gallons in volume (158.987 liters). Although not related to the metric system, this unit of measure has become the primary international standard. One barrel of oil contains approximately 5.71 GJ, which can also be referred to as 1 **barrel of oil equivalent**. For purposes of international trade in oil, the primary measure used expresses production in **millions of barrels per day**, which helps us better understand how much oil is being pumped and shipped around the world.

Image 1.6 Oil continues to be measured via an arcane volumetric measure, the barrel. An oil barrel is 42 gallons.
Wasan Soisong / Shutterstock

Current Energy Sources

Today, the majority of humanity's energy needs are met using **nonrenewable energy resources**. Nonrenewable energy resources are energy resources that, once converted for use by people, are gone forever. Their energy conversion is a one-off trade that often comes with vast consequences. Four primary sources of nonrenewable energy are in use today: coal, oil, natural gas, and nuclear energy. We also derive energy from a series of **renewable energy resources**, which derive energy from ongoing natural processes and cannot reasonably be exhausted. This natural replenishment allows for continual use with less concern for resource depletion. Common renewable energy resources used today include biomass, hydropower, wind, geothermal, and solar energy.

Almost all forms of energy we consume originally come from **solar energy**, transmitted as light or **radiant energy**. The constant arrival of light and heat energy from the sun to the surface of the earth provides the source for such varied forms of energy as the **fossil fuels** (coal, oil, and natural gas), wind, solar, and hydropower. Fossil fuel energy is sunlight stored by various plants and unicellular microbes through photosynthesis, which then became trapped deep under the earth's surface and transformed into an energy resource. Only later have these resources been harvested and converted by humans for various uses. The renewable energy resources derived from the sun are harvested from natural solar-driven processes that impact the earth's surface, hydrosphere, and atmosphere. Solar radiation drives the wind, rainfall, and other processes that are then harvested through renewable technologies.

Only nuclear and geothermal energy (with the exception of a few minor technologies such as tidal energy) do not derive their energy from the sun. Nuclear energy utilizes uranium mined from the earth's crust, which dates to the formation of the planet. Geothermal energy relies on the heat energy that is produced from the massive pressure under the earth's crust, as well as the heat resulting from radioactive decay of elements within the crust.

Electrical Energy

Electricity is probably the most important single secondary energy source. It is secondary because we use other energy resources to create electricity, which is then transmitted and consumed in its pure form. Electricity is the flow of an electrical charge, or electrical power. Since electricity is produced from both renewable and nonrenewable resources, it should not be thought of as either renewable or nonrenewable. It is hard to fathom life without electricity because it powers so much that we use every day, but electricity has only been directly used by people for a little over one hundred years.

Electricity has been a part of our world since long before human beings learned how to manipulate it for our purposes. **Static electricity**, produced through friction, causes electrons to jump from one atom to another. A magnetic field is another phenomenon through which we saw electricity prior to its industrial production because the arraying of forces from the magnets is related to the electrical charge of the electrons in the magnets themselves (Image 1.7). For the widespread use of electrical power, we use magnets in an **electrical generator**. The British physicist Michael Faraday discovered that moving a magnet inside a coil of wire led to the formation of an electrical current within the wire. In a modern electrical generation facility, a coil of wires is surrounded by an electromagnet, and physical energy is used to turn the coil rapidly between the magnets, leading to the generation of power in each wire of the coil, which is then

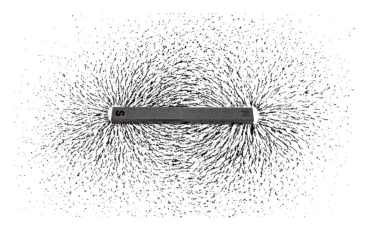

Image 1.7 Magnetic fields demonstrate properties similar to electric fields, and magnets are used in the generation of electricity.
ImageDB / Shutterstock

consolidated into a larger **electrical current**. A variety of sources are harnessed to turn these coils of wire, including **steam** turbines, wind and water turbines, and internal combustion engines, to name just a few (Image 1.8).

In the United States and most developed countries, the majority of electricity is produced by steam turbines. In these systems, an energy input is used to boil water in a boiler, which produces steam that is then pumped through a turbine system consisting of a series of fans, much like those in an airline jet engine (Image 1.9).

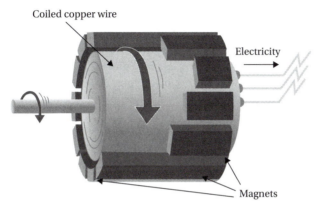

Image 1.8 Most electricity we consume is produced in generators that use large coils of wires that rotate inside electromagnets to produce a flow of electrons.

Courtesy of the U.S. Energy Information Administration

Image 1.9 Steam turbines, such as those used in coal, natural gas, and nuclear plants, have large fan blades that are turned by steam from a boiler flowing through them, similar to how jet engine blades turn.

Monty Rakusen / Cultura / Science Photo Library

These fan blades turn a rotor that is then connected to the coil of wires inside the generator, which spins to produce the electrical current. The steam is then condensed back into liquid water and recycled through the process. We will look at the specific means by which this is done as we investigate different energy resources throughout the book.

Once the current is produced, it must be transmitted to its end uses. The electricity is run through a transformer that steps up the voltage of the electricity for its journey through high-capacity transmission lines. Further transformer facilities located near end users lower the voltage for use again on the other end. The electricity is then transmitted through normal power lines to its end users in homes, offices, stores, and factories (Image 1.10). This process is efficient, but small losses of electricity occur along the transmission route, which we refer to as **transmission loss**. Approximately 5 to 6 percent of all electricity that departs a power plant is lost in transmission, as electrons are shed from the high-capacity power lines. This loss is mitigated by locating electrical generation facilities close to end users, because longer distances increase transmission loss.

At these end uses, we can turn the electricity on or off using circuits. Electricity moves only through a closed **circuit**; without a complete path, the electrons are not able to move and remain in balance at the same time. The flipping of switches closes the circuit, allowing electricity, and thereby power, to flow through to the end use device, from something as small as a clock or speaker to large appliances, supercomputers, and large industrial machines.

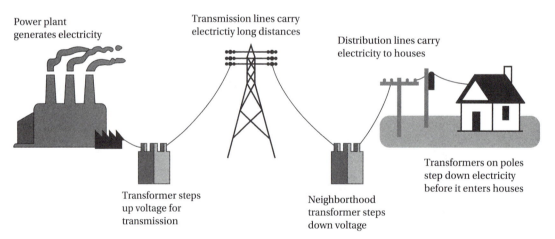

Power plant
generates electricity

Transmission lines carry
electrictiy long distances

Distribution lines carry
electricity to houses

Transformer steps
up voltage for
transmission

Neighborthood
transformer steps
down voltage

Transformers on poles
step down electricity
before it enters houses

Image 1.10 This model demonstrates the process by which electricity moves from a power plant to final consumers of energy.

Courtesy of the U.S. Energy Information Administration

Electricity is a basic natural phenomenon, although the steps required to harness it have only recently been employed by humanity. Two scientists revolutionized our use and understanding of electricity to make possible its widespread use, replacing other less efficient, and in many cases more dangerous, predecessors to provide for basic human needs. Thomas Edison's invention of the light bulb made possible the elimination of kerosene lamps and candles as the primary means of lighting homes, businesses, and streets. Nicola Tesla's experiments that led to developing **alternating current** made it possible to transmit electricity over much longer distances without significant transmission loss.

Energy Efficiency

One of the great challenges we face in our energy system is its lack of efficiency. This challenge is encountered at the production, transmission, and consumption phases of the system, and a more sustainable future will require humanity to overcome the gross inefficiencies that mark our current system.

At the production end, inefficiencies abound, starting with the harvesting of different resources. We will explore these inefficiencies in depth in the following chapters. At a general level, however, we leave huge amounts of fossil fuels in the ground that cannot be harvested, and our alternative energy systems also are unable to convert large amounts of the potential energy they encounter into usable energy. The production of electricity is inefficient in most power plants. While some technologies allow for better generation efficiency, many traditional power plants (such as coal power facilities) only convert around 35 percent of the total consumed energy into electricity. Improving this efficiency may be one of the keys to dramatically curtailing humanity's overall energy consumption.

For electricity, the inefficiencies continue in dramatic fashion after initial production. Electricity moves from the power plant out into the grid, where large amounts of electricity are lost during transmission.

Further inefficiencies result from the varying demands for energy at different times of the day and year. Most large-scale power plants, especially those running on coal or nuclear power, take a long time to fire up and to scale back. The system must always produce enough power to meet the demand at that time, thus causing the plants to constantly run above demand. This means that many large power stations are producing significant amounts of electricity beyond what is being used at the time because of the inability to fire up extra generation fast enough to meet varying demands at different times of the day (Image 1.11). When this extra generation fails, rolling blackouts occur, cutting off electricity to consumers. While they are rare in the developed world, partial blackouts are a daily occurrence in some less developed regions because of a lack of fuel or infrastructure.

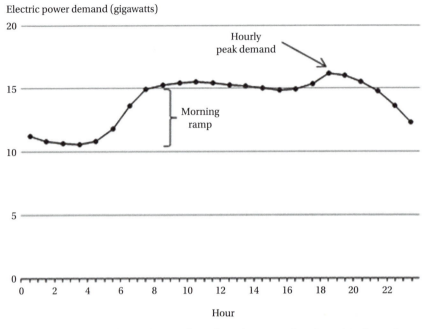

Electric load curve: New England, 10/22/2010

Electric power demand (gigawatts)

Image 1.11 Electricity demand varies throughout the twenty-four-hour day, depending on when lights are needed, appliances are run, and heating and cooling systems are running at various amounts. This variance differs throughout the year and varies depending on location and the weather on any given day.

Courtesy of the U.S. Energy Information Administration

Energy Storage

As the previous discussion of variable energy demand demonstrates, generating the right amount of electricity at the right time is a major factor in running a power grid. Generating excess electricity wastes fuel (in fossil and nuclear fuel systems), and when electricity supply falls short, blackouts cause serious difficulties and impose major costs on society and the economy. With renewable energy generation, variability is a much more serious problem because the amount of sunlight, wind, and water flow is mostly beyond our control. Energy storage, particularly **grid storage**, allows electricity from periods of excess generation to be stored for use later when the grid is no longer generating enough to meet demand. On a small scale, batteries and other storage technology have been used for years to ensure constant electrical supply in case of disruptions, but grid-level storage solutions are becoming more and more critical as highly variable renewable electricity gains a larger share of our total energy system.

One of the most reliable methods of grid energy storage is **pumped-storage hydroelectric**, which uses the same principles as conventional hydroelectricity (Image 1.12). In periods of excess generation, water from one reservoir is pumped uphill to

Image 1.12 This pumped-storage system in Colorado provides a means to store electricity for use later in an efficient manner.

Jim West / Science Photo Library

another storage reservoir, converting the electrical energy to potential energy. When the energy needs to be converted back to electricity, the water from the top reservoir is then allowed to flow through turbines going downhill, using gravity to generate electricity. Because of energy losses in pumping water and converting it back and forth, pumped storage is usually between 70 and 80 percent efficient, so 1 kWh of electricity stored would only return 0.7 to 0.8 kWh back into the grid. This system is relatively efficient in the context of energy storage methods. Because of the simplicity of the technology and infrastructure needed for pumped storage, it has been the overwhelming favorite worldwide, representing nearly 99 percent of all grid energy storage. Significant barriers include cost (building multiple reservoir systems can be expensive) and siting, because a pumped-storage system requires abundant water resources and enough natural variation in elevation to place the two reservoirs at the appropriate heights.

Battery storage has been applied on a small scale for many uses, such as cellular phones and laptop computers, but battery technology has only recently advanced sufficiently to potentially allow grid-level storage. Grid battery storage could directly store electricity in a chemical battery (Image 1.13), which has the benefit of ease of use almost anywhere, unlike pumped hydroelectric. Storage from large renewable facilities could be concentrated in large battery banks onsite or could be distributed throughout the power grid to ensure reliability and minimize disruptions. Smaller storage batteries can be used in conjunction with technologies such as rooftop solar to store generated electricity on-site at the homes or businesses that use it, and these batteries could also store electricity from the grid.

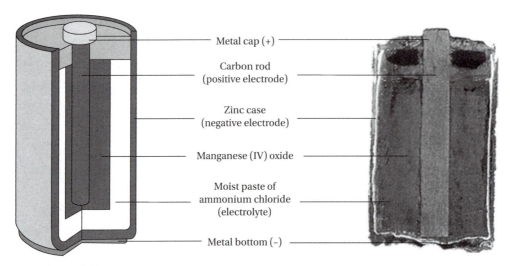

Metal cap (+)

Carbon rod
(positive electrode)

Zinc case
(negative electrode)

Manganese (IV) oxide

Moist paste of
ammonium chloride
(electrolyte)

Metal bottom (–)

Image 1.13

As electric cars become more common, the concept of **vehicle-to-grid** storage has been proposed, in which the batteries of electric cars that are plugged in while not in use can be used as grid batteries. This method will only work if electric cars begin to be adopted on a large scale, but has the benefit of combining several clean technologies to solve multiple challenges within our energy system.

Batteries are expensive, but as costs continue to drop and more early adopters install them to store renewable electricity, a rudimentary storage grid will begin to grow. If grid storage can keep pace with growth in renewable generation, the two systems could theoretically be scaled up together to make a grid run entirely on renewable sources. This system would require further advances in battery technology (potentially supplemented by other storage methods) and continued reductions in battery cost, but the current pace of battery technology at least hints that this may be possible in the coming years.

Other methods of energy storage exist, including the following:

- Compressed-air storage—gas is compressed with excess electric power and stored within large containers or underground reservoirs, and the gas can be used to run turbines to convert back into electricity.
- Hydrogen storage—a type of chemical storage in which electricity is used to hydrolyze water and capture hydrogen gas. The hydrogen can then be burned to generate electricity, as in a conventional thermal turbine.
- Thermal storage—heat from solar power or waste heat from thermal generation can be stored in a liquid (usually a type of oil or molten salt), which can then be used to run a thermal generator. This type of storage works well for generation methods that already produce large amounts of heat, but the heat dissipates the longer it is stored, so efficiency is highly variable.

At present, worldwide energy storage is mostly accounted for by pumped hydroelectric (with around 130 GW of capacity). As renewable energy generation expands, the pressure to develop affordable and practical grid storage increases as well. By combining multiple storage methods, as well as continuing to develop energy storage technology, distributed grid storage may become much more common in the near future. If industrialized countries continue their push to reduce emissions and rely on renewable electricity in even greater proportions, grid storage will have to become a major part of our global energy system.

REVIEW QUESTIONS

1. When exploring energy resources, in what forms of energy do these resources occur?
2. What is the difference between a kilowatt and a kilowatt-hour? Why do we need both units of measure?
3. Is electricity a primary or secondary energy resource and why?
4. Why must we keep electricity production relatively close to electricity consumption locations?
5. Why can't we utilize pumped storage everywhere?

Image 2.1 ArtisticPhoto / Shutterstock

CHAPTER 2

The Energy Basket

Rather than relying on a single source for our energy needs, humanity relies on a diverse "basket" of resources. These resources include the earliest fuels of wood and animal fat, through moving water to grind grains and power machines, to coal and oil (and modern technologies such as solar panels and nuclear fission). Together, these resources make up the global energy economy. Some resources are used close to where they are harvested, while others travel thousands of miles prior to their consumption. When we analyze the energy economy at different scales—be it that of a region, country, province, or city, among others—we can look at the different assortment of resources to understand how a place fits into the global energy picture.

Through this analysis, we can look at not only how much energy is consumed and which resources provide it, but also the sectors of the economy that consume energy and where energy is traded around the globe. This gives us a nuanced understanding of the varying energy systems we see throughout the world and provides a basis for the comparison of different resources and their use in the global energy system.

In this chapter, we identify the major energy resources that make up the global energy basket, along with where energy consumption occurs globally, and how we produce energy and trade it at different scales. We will then identify the energy flow and understand what resources make up the United States' energy basket and how that mix of resources and consumption has changed over the past two hundred years.

Energy Sources and Trade

Our energy resources can be categorized into two broad groups: renewable and nonrenewable (Image 2.2). A great variety of resources constitute both the renewable and the nonrenewable categories, each of which is discussed in the following chapters.

Nonrenewable energy resources make up the vast majority of what humanity extracts and consumes today. These resources are called nonrenewable because they do not replenish within our lifetime or, indeed, within hundreds or even thousands of years. The **fossil fuels**—coal, oil, and natural gas—are derived from microorganisms and plants that lived millions of years ago. More fossil fuels are always being formed by geological processes deep underground; however, these processes take hundreds of thousands to millions of years, which in effect makes them nonrenewable, although the earth will produce more of these resources eventually. The other major nonrenewable energy resource is nuclear, which is derived from radioactive elements such as uranium, plutonium, or thorium. Because these sources are pure elemental forms, no more can be created, so our supply is limited to what has been present since the formation of the earth.

The other group consists of **renewable energy resources**. These resources are constantly being replenished and do not run the risk of running out. The primary renewables used today include hydropower, wind, solar, biomass, and geothermal energy. Other than geothermal, these resources all derive their energy from solar radiation that has recently come to the earth. Other than nuclear and geothermal, all other major energy resources, whether renewable or nonrenewable, can trace their energy to prehistoric or current radiation from the sun (Image 2.3).

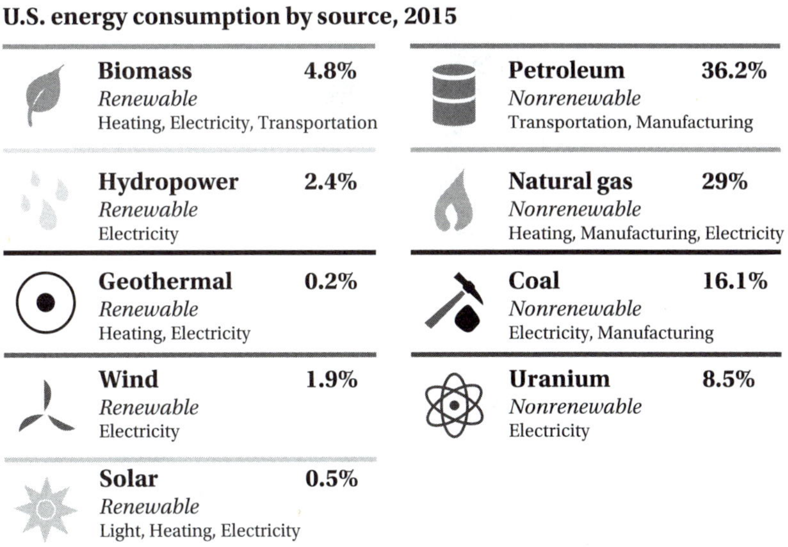

U.S. energy consumption by source, 2015

Biomass 4.8% *Renewable* Heating, Electricity, Transportation			**Petroleum** 36.2% *Nonrenewable* Transportation, Manufacturing	
Hydropower 2.4% *Renewable* Electricity			**Natural gas** 29% *Nonrenewable* Heating, Manufacturing, Electricity	
Geothermal 0.2% *Renewable* Heating, Electricity			**Coal** 16.1% *Nonrenewable* Electricity, Manufacturing	
Wind 1.9% *Renewable* Electricity			**Uranium** 8.5% *Nonrenewable* Electricity	
Solar 0.5% *Renewable* Light, Heating, Electricity				

Image 2.2 The majority of America's current energy basket is nonrenewable, and renewables are a growing source of our energy.

Courtesy of the U.S. Energy Information Administration

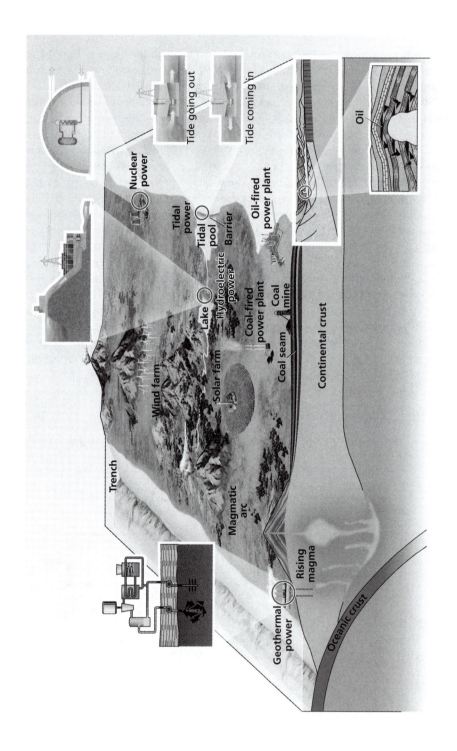

Image 2.3 Geothermal, nuclear, and tidal power are not derived from the sun, while all other energy resources on earth are initially derived from solar energy.

All energy resources are distributed unevenly around the globe, giving certain geographic locations natural advantages or disadvantages. Unique geological conditions are responsible for the Middle Eastern prominence in the oil industry, a wealth of coal and natural gas in the American West, and particularly rich uranium deposits in Australia, for example. Because fossil fuels require specific natural conditions to form over millions of years, only certain places have economically viable deposits. As for renewable resources, wind and solar radiation occur at varying intensities and durations in different latitudes and climates, and precipitation and water flow for hydropower are also concentrated in specific locales. Consequently, certain resources can only be exploited in specific geographic locations, some of which are located far from the human populations that use them.

A complex international web has developed to move resources and produced energy from one geographic location to another. Nonrenewable resources are mined or extracted and then transported to other locations for consumption, often crossing international borders in the process. Ships bring coal, oil, liquefied natural gas, and nuclear fuels to faraway locations for processing and consumption (Image 2.4). Oil and gas move via pipelines across international borders as well. Electricity can be moved thousands of miles efficiently by use of high-voltage transmission lines (Image 2.5). Renewable energy is often converted to electricity at the location where the resource is harvested and transmitted to consumers. Biomass can be burned at home or converted into liquid fuels or pellets and shipped to consumers.

Electricity is sold across international borders in many places, and large regional markets for electricity have developed in many parts of the world, especially in Europe and South America. France is the world's leader, exporting large amounts of nuclear-generated electricity to other European countries, while Paraguay, the world's fourth largest exporter, directs hydroelectric-generated electricity to its larger neighbors, Brazil and Argentina (Table 2.1). As renewable electricity grows in prominence, it will likely become a much larger part of the global electricity trade. Places with a geographic advantage in generating

Image 2.4 Large tankers are just one way that we move energy resources long distances from source to market.

Ian Tragen / Shutterstock

Image 2.5 Even electricity can be exported to neighboring countries over high-tension power lines.
Konstantinks / Shutterstock

Table 2.1 Top International Electricity Exporters, 2014 (estimates)

Country	*Electricity exported (billion kilowatt-hours)*
1. France	75.0
2. Germany	74.0
3. Canada	58.4
4. Paraguay	41.0
5. Switzerland	34.0

Source: CIA World Factbook.

renewable electricity such as hydropower, solar, and wind will be able to export electricity to neighboring countries and regions that lack the same access to renewable resources.

World Primary Energy Consumption

The global economy is sustained by massive inputs of energy from all available sources, and every economic sector requires a constant energy supply to function. The International Energy Agency expects global energy demand to grow by 37 percent from 2015 to 2040, or about 1.5 percent per year from a baseline of over 150,000 TWh. Over the past two decades, global energy consumption grew at a rate of about 2 percent per year.

Not every country has experienced rising energy consumption. Breakthroughs in energy efficiency have largely been responsible for the slowdown in energy demand growth, with demand remaining flat in many industrialized countries, despite population increases. But with growing populations and a fast-emerging middle class in many developing countries, global energy needs will continue to rise.

Currently, the global energy basket is dominated by fossil fuels, which account for approximately 82 percent of global energy consumption, according to the International Energy Agency (Image 2.6). Oil occupies the largest share of the basket, accounting for over 31 percent of global consumption, followed by coal at 29 percent and natural gas at about 21 percent. The largest renewable source of energy globally is biofuel, at 10 percent of total consumption; most of this use, however, is direct combustion of firewood or plant products for heating and cooking in less developed countries. Nuclear power accounts for less than 5 percent of the global basket, with hydropower being the only other source that holds a significant share, amounting to a little over 2 percent (Table 2.2).

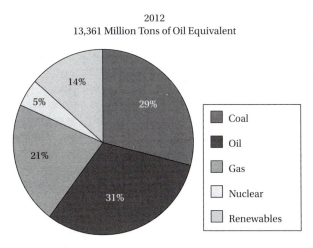

2012
13,361 Million Tons of Oil Equivalent

- Coal
- Oil
- Gas
- Nuclear
- Renewables

Image 2.6 Coal, oil, and natural gas dominate current global energy consumption.

Information courtesy of the IEA 2014 World Energy Outlook

Table 2.2 World Primary Energy Demand by Fuel (million tons of oil equivalent)

	2012	2020 estimate with current policies	2040 estimate with current policies
Coal	3,879	4,457	5,860
Oil	4,194	4,584	5,337
Natural gas	2,844	3,215	4,742
Nuclear	642	838	1,005
Hydroelectric	316	383	504
Bioenergy[a]	1,344	1,551	1,933
Other renewables	142	289	658
Total	13,361	15,317	20,039
Fossil fuel share (%)	82	80	80
Non-OECD share (%)	60	63	70

Note. OECD, Organisation for Economic Co-operation and Development.

[a]Includes traditional and modern uses of biomass fuels.

Source: International Energy Agency 2014 World Energy Outlook.

This consumption of energy is in no way evenly distributed among the world's population, with a vast gap in energy consumption between developed countries and developing countries, and there are also major differences between countries in each category. The **Organisation for Economic Co-operation and Development**, mostly associated with the developed countries, accounts for about 40 percent of global energy consumption, but less than one billion of the earth's population of over seven billion people.

On a gross scale, China is now the world's number one energy consumer, followed by the United States. On a per capita basis, China lags far behind members of the Organisation for Economic Co-operation and Development, but is far ahead of many developing countries. India, the world's second most populous country, consumes less energy than not only China and the United States, but also Russia and Japan, whose populations are less than 10 percent of India's. Just among this small number of countries, vast disparities in energy consumption manifest.

The starkest contrasts in the global energy basket can be seen by comparing energy consumption in sub-Saharan Africa, the earth's poorest region, with that of the rest of the world. Over 60 percent of the region's population does not even have access to a regular supply of electricity. Many sub-Saharan countries consume firewood and other local biofuel sources as their primary means of harvesting energy, with more rudimentary electricity grids confined to larger cities (Image 2.7).

As we investigate the different forms of energy resources, it is important to understand the geographic disparities in reserves, production, and total energy consumption. By looking at the total energy system, it is possible to understand how energy consumption impacts the politics, society, and environment of differing cities, regions, and countries around the world.

Image 2.7 In certain parts of the world, such as this scene in the Kalahari Desert in Africa, people still rely on firewood or other locally sourced organic materials as their primary source of energy.

Lucian Coman / Shutterstock

VIGNETTE

The Industrial Revolution and the Change to an Energy-Intensive Economy

Prior to the start of the **Industrial Revolution**, humanity consumed far less energy overall and per capita. Biomass, primarily firewood, was the primary source of energy worldwide, and human and animal muscles were the primary way of getting anything done. The game-changing technology that launched us into the energy-intensive economy and moved us from muscles to machines was the steam engine. Although theoretical steam engines were proposed as far back as ancient Alexandria, Egypt, it was the engine developed by James Watt in Scotland that revolutionized our use of energy and launched the Industrial Revolution.

Prior to Watt, the first **steam engines** were developed in England for use in coal mines by the 1760s (coal was already in use as a heating fuel in parts of Great Britain as a result of widespread deforestation) (Image 2.8). But it was Watt's inventions in the 1760s that launched a wholesale expansion of the use of fossil fuel power in the broader economy.

As a result of Watt's numerous advances in the steam engine, the uses for coal in various areas of the economy grew rapidly. This growth began with the development of stationary engines that could be used in mines and factories to run large machines. Soon thereafter, the first steam engines were developed for transportation. Thus, it was in the industrial and transportation portions of the economy where the Industrial Revolution first took hold and where humankind first escaped the bounds of what could be accomplished solely using biomass energy.

The Industrial Revolution dramatically altered the face of society. Being able to harness such large amounts of energy to accomplish basic tasks allowed for a massive shift from subsistence agricultural life to an urban/industrial way of life. Millions of people moved from farms to cities seeking jobs in factories (Image 2.9). Standards of living began to rise as goods that previously were produced by hand could now be made with large machines, at first bringing down the price of goods like cloth and simple home devices and later producing machines that individuals could use.

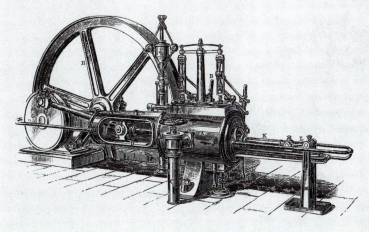

Image 2.8 The steam engine, powered by a coal-fired boiler, was the primary invention that led to the Industrial Revolution and its concurrent massive increase in energy consumption.

Hein Nouwens / Shutterstock

continued

continued

Image 2.9 Steam power was transferred to run smaller machines in factories, such as the textile machines in this English factory.
Neveshkin Nikolay / Shutterstock

Great Britain, the center of the technological developments in steam engines, was also home to large coal reserves and thus became the first to transition to a modern industrial economy. This in turn impacted its political position, as the harnessing of all this energy created the wealth and power that led to Britain becoming the nineteenth century's greatest geopolitical power. The United States and Germany, also home to large coal reserves, followed Britain into the Industrial Revolution, and slowly, other states in Europe and later Japan industrialized, helping to crystallize the global political power structures of the nineteenth and twentieth centuries.

By the start of the nineteenth century, the Age of Steam was in full force. The steam-powered locomotive improved the speed and power of land transportation so that people could move goods and people far faster overland. The steam ship removed our need for wind as the primary source of energy for oceanic transit, expediting travel overseas. This in turn spurred a massive population shift as many impoverished Europeans from economically marginal places in the south and east of Europe migrated not only to cities, but also to the booming industrial centers of the United States.

The Industrial Revolution, as an energy-driven revolution for humankind's way of life, led to massive changes not only in the economy, but also in our politics and society. It also initiated a process that continues today, known as the **Anthropocene era**, where humans became the single most important component in changing the global ecosystem, especially its climate. These intertwining processes of industrialization and intense energy consumption continue to spread as the least developed parts of our planet are now in the process of this transition.

The Energy Basket of the United States

The United States has the most diverse energy economy in the world, by virtue of its significant consumption and production of energy resources. With a large population, geographic diversity of climates, and the world's biggest economy, the United States consumes vast amounts of resources and makes for a good case study of both energy resource production and consumption. Every major energy resource is both harvested or exploited within the United States and consumed by the American energy economy. This section will examine the overall trends to provide a baseline that can then be compared to the specific energy resources throughout the remainder of the book.

The United States domestically produces or exploits every major energy resource, with vast amounts of fossil fuels located under American soil or in offshore deposits. Uranium is also mined in the United States, although imports have displaced much of the domestic production since prices dropped in the 1980s. The United States also has large resources of hydropower, wind, solar, and biofuels, with lesser amounts of geothermal energy.

As illustrated in Image 2.10, the United States consumes large amounts of every energy resource. Petroleum or oil consumption comprises the single largest provider of energy in the United States, consisting of just over one-third of total energy consumption. Natural gas has overtaken coal recently as the number two source, with coal placing third. Together, the three fossil fuels account for over two-thirds of the total energy

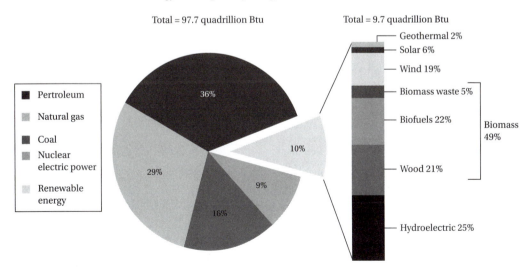

U.S. energy consumption by energy source, 2015

Total = 97.7 quadrillion Btu   Total = 9.7 quadrillion Btu

Legend:
- Pertroleum
- Natural gas
- Coal
- Nuclear electric power
- Renewable energy

Geothermal 2%
Solar 6%
Wind 19%
Biomass waste 5%
Biofuels 22%
Wood 21%
Hydroelectric 25%
Biomass 49%

36%
29%
9%
16%
10%

Image 2.10 The majority of U.S. energy comes from nonrenewable sources. Btu, British thermal unit. Courtesy of the U.S. Energy Information Administration

resources consumed in the United States. The remainder of American energy con-sumption is divided between nuclear power and all renewables, which together only make up about 10 percent of total energy consumed in the United States.

Not all resources in the U.S. energy basket go to the same sectors, nor do all sectors consume the same amount of energy (Image 2.11). Generation of electricity is by far the largest single use of energy resources in the United States, and our massive electric output is an essential component of all other economic sectors, from industry to home usage. While electric generation claims the largest total energy input, the most consumed resource, petroleum, is largely used in the transportation sector, with 71 percent of all petroleum converted into fuel (and accounting for 92 percent of all energy consumed by transportation needs). The differences in consumption of

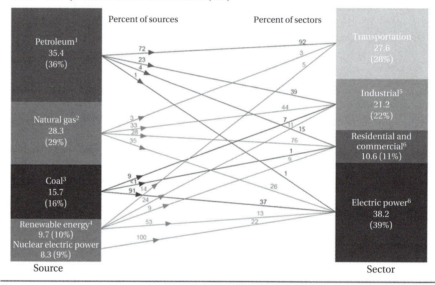

U.S. primary energy consumption by source and sector, 2015
Total = 97.7 quadrillion British thermal units (Btu)

[1] Does not include biofuels that have been blended with petroleum—biofuels are included in "Renewable Energy."
[2] Excludes supplemental gaseous fuels.
[3] Includes less than –0.02 quadrillion Btu of coal coke net imports.
[4] Conventional hydroelectric power, geothermal, solar/photovoltaic, wind, and biomass.
[5] Includes industrial combined–heat–and–power (CHP) and industrial electricity–only plants.
[6] Includes commercial combined–heat–and–power (CHP) and commercial electricity–only plants.
[7] Electricity–only and combined–heat–and–power (CHP) plants whose primary business is to sell electricity, or electricity and heat, to the public. Includes 0.2 quadrillion Btu of electricity net imports not shown under "Source."

Notes: Primary energy in the form that it is first accounted for in a statistical energy balance, before any transformation to secondary or tertiary forms of energy (for example, coal is used to generate electricity). Sum of components may not equal total due to independent rounding.

Image 2.11 The relationship between energy resources and the economic sectors they power.
Courtesy of the U.S. Energy Information Administration

various resources among different end uses demonstrate that these resources cannot necessarily be directly substituted for one another. For example, oil for transportation cannot be replaced by renewable electricity generation without a shift to electric vehicles. The electricity sector is more flexible than others because nearly all energy resources can be employed in power generation, although efficiencies and costs can vary considerably. Oil, however, makes up only 1 percent of national electricity generation in the United States, while coal, natural gas, and nuclear energy all play major roles (Image 2.12).

Over time, we expect the energy basket in the United States to shift in response to changing prices for resources, changing rules on pollution, and other regulations and technological advances. Based on current trends, we anticipate that the electricity market will continue to grow in the United States and will also remain diverse compared to that of most other countries in the world (Image 2.13). The U.S. Department of Energy estimates that natural gas will be the leading fuel for additional growth, followed by renewables, while coal and nuclear use will continue at levels similar to those in use today. However, all such estimates must be regarded with caution because changes in the energy basket can shift quickly. In less than a decade, cheaper natural gas prices have led to massive growth in natural gas consumption, primarily displacing coal, none of which was foreseen in contemporary resource assessments.

As we explore the various resources, it is always helpful to think about their use as part of the broader energy structure. Keep in mind the different sectors in which the resource plays a role and how that could change in the future, as well as the other resources it competes against for its spot in the overall energy picture.

Energy consumption (Reference case)
Quadrillion british thermal units

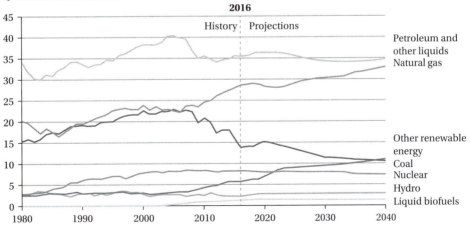

Image 2.12 While oil consumption is forecast to remain steady, the U.S. Department of Energy estimates that natural gas and renewables will grow fastest to meet growing demands by 2040. It is important to note that these estimates could change depending on many factors over the coming decades.

Courtesy of the U.S. Energy Information Administration

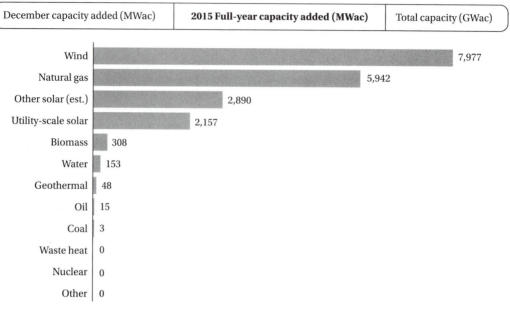

| December capacity added (MWac) | 2015 Full-year capacity added (MWac) | Total capacity (GWac) |

Wind — 7,977
Natural gas — 5,942
Other solar (est.) — 2,890
Utility-scale solar — 2,157
Biomass — 308
Water — 153
Geothermal — 48
Oil — 15
Coal — 3
Waste heat — 0
Nuclear — 0
Other — 0

Image 2.13 Most new electric capacity coming online in the United States is from renewable sources and natural gas.

"New US Electricity Generation Capacity (Jan-Dec 2015)" / Zachary Shahan / CleanTechnica / 2016

History of Energy Consumption in the United States

Today's energy basket looks different from that of previous eras. The energy picture has changed dramatically over time as new technologies have been developed and total energy consumption has skyrocketed. In addition to the basket of today, it is essential to have a basic understanding of the changes seen over time in energy. It must be kept in mind that the United States was one of the first countries to develop a diverse energy mix, along with other societies that industrialized earlier. Many other countries have followed a similar trajectory in more recent times, as they have moved along the continuum of becoming developed economies. Distinct geographies in some countries dictate a less diverse overall energy picture than that of the United States.

When the United States became independent, wood served as the primary source of energy. At that time, sailing ships were powered by wind and some water power was used to mill grains and for other simple industrial processes. Other than these exceptions, almost all energy consumed came from the burning of wood, which heated homes, cooked food, and boiled water for cleaning. Wood dominated because it was locally available, easy to transport, and could be consumed when needed. Industry was small-scale and local, and transportation was mostly accomplished on foot or

History of energy consumption in the United States, 1775-2009

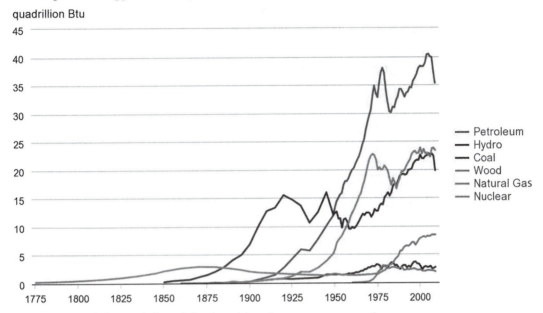

Image 2.14 Wood, then coal, then oil dominated American energy consumption.

with horses. Indeed, animals played an important role in energy production, as not only horses, but also oxen, mules, and donkeys provided means to transport goods and people and to power simple machines.

The 1800s brought major changes to energy production and consumption in the United States and in other emerging industrial countries such as the United Kingdom, Germany, and later Japan. The first resource to emerge as a major competitor to wood was coal. By the mid-1800s, coal began to be used as an important energy resource in the United States, but it had emerged in the United Kingdom a couple centuries earlier as a major resource. Just like wood, coal was easy to transport and could be consumed on demand. Coal was first used in Pennsylvania and New York, near where the first economically viable deposits were found. With the rise of industry and the invention of electric generation methods for widespread use, coal consumption skyrocketed and took over as the largest energy resource in the United States by the 1880s.

Oil production in the United States began in the 1860s (Image 2.15) and overtook coal as the nation's primary energy resource by the 1950s. Oil was used first for home lighting and heating and only later for transportation and electricity generation. Although oil consumption varied and saw declines in the later 1900s, it has continued to remain the dominant energy resource in the U.S. basket, although its use has moved primarily to transportation, with only smaller amounts going to other sectors over time.

Image 2.15 Oil production started in the United States in Pennsylvania, one of the first places where oil was produced on a commercial scale.
Morphart Creation / Shutterstock

Natural gas emerged next as a U.S. resource, starting in the later 1870s. It heated homes and offices and later fueled electric power plants. By the 1950s, natural gas overtook coal as the number two energy resource in the United States, but high gas prices and low coal prices led to fluctuations, and natural gas has only recently regained the number two position in terms of American energy consumption.

Other resources have also played an important role over the history of American energy consumption. Hydropower has been in use since the country's founding; however, it only emerged as a major resource in the early 1900s with the advent of hydroelectric power plants. Large dams were built in the first half of the century, many during the Great Depression, and hydropower continues to serve as one of the most important renewable resources in the country. Nuclear power emerged in the 1950s and took off quickly, but has stagnated since 1979, when the Three Mile Island nuclear accident led to an abrupt halt in the growth of nuclear power in the United States. Only in the most recent decades have other renewable resources begun to play an important role in the American energy economy (Image 2.16). Forecasting the future of something as complicated as a total energy system is difficult to do with any precision, but many analysts expect the trend of growth in renewables to continue, possibly even leading to major declines in usage of the traditionally dominant fossil fuels. If technology can advance these resources sufficiently, it will be possible for the fundamental structure of the world energy basket to change in ways as dramatic as the shift to coal and oil brought about by the Industrial Revolution. In the coming chapters, we will analyze each of these resources with the aim of developing possible trajectories for the future of energy in the United States and the rest of the world.

Image 2.16 Renewable energy is just now becoming an important part of the United States' energy basket.

Jesus Keller / Shutterstock

REVIEW QUESTIONS

1. What is the original source of almost all our energy resources, renewable and nonrenewable? What are the exceptions?
2. Although not a source of energy in and of itself, what is the dominant form of energy consumed by people today?
3. While we think of the trade in commodities like oil, how does the global market for electricity work differently? Why are the dominant players different?
4. Global energy consumption is dominated by nonrenewable resources. Which renewable resources already play a major role in global energy consumption and why?
5. How can the Industrial Revolution be seen primarily as an energy revolution?
6. Why can the United States be considered to have the most diverse energy economy? Which resource is fast taking on coal as the dominant resource?

Image 3.1 Dyzio / Shutterstock

CHAPTER 3

Coal

Coal: dirty, cheap, ubiquitous. It emerged as the leading fuel of the Industrial Revolution and remains one of our most important sources of energy. Other than perhaps oil, no energy resource provokes as much controversy today as coal and the world's continuing reliance on it. Coal may be controversial today, but its status as a primary part of the global energy system dates back far into the past, and its uses have varied significantly over time.

Coal has been used in a variety of ways for thousands of years, with archaeological examples in China dating to around 3500 BCE. In some ancient societies, the shiny variety of lignite known as jet was used to make jewelry (from which the term *jet black* is derived) (Image 3.2). Roman Britain was the first place where coal was used widely as an energy source. Its many purposes included the heating of forts and homes, the smelting of metal, and a burnt offering to honor the gods at temples. Despite the widespread use of coal in Britain, other parts of the Roman world did not begin to use it in significant quantities until hundreds or thousands of years later, other than importing a small amount of jet for jewelry. During the Middle Ages, coal was used both for home heating and for forging by blacksmiths in Britain, especially in the coal-rich midlands; by the seventeenth century, coal was the main fuel used in England. As a result of Europe's rapid deforestation, other countries also began to turn to coal as a fuel source during the Renaissance.

The great advent of coal as a global energy resource began with the Industrial Revolution. The development of the coal-powered steam engine in the 1770s led to the emergence of large-scale manufacturing, railroads, and steamships (Image 3.3). Together,

Image 3.2 Jewelry made with jet, which is coal. Courtesy of Detlef Thomas

Detlef Thomas / Wikimedia Commons / CC-BY-SA-3.0

Image 3.3 Early steam engines used coal as fuel, driving the Industrial Revolution.

The first steam engine designed and built in the United States, by Oliver Evans, of Philadelphia, Pa., 1801 / Drawing by Thos. Arnold McKibbin / Library of Congress

these coal-fired technologies revolutionized industry and transportation, broadening coal's applications beyond heating and metallurgy. The newly discovered uses of electricity further cemented coal's position as a key resource with the development of thermal electrical generating plants at the end of the nineteenth century.

By the start of the twentieth century, coal was the most common fuel in the industrializing parts of the world. Although petroleum products eventually eclipsed coal as the primary fuel in the global energy system, coal remains one of the main sources of today's energy economy, primarily as a fuel for electrical generation in both developing and highly developed countries. Additionally, coal's high energy content ensures its continuing importance in industrial production, especially metallurgy.

Coal Formation

Coal, along with all other fossil fuels, derives its original energy from the sun and can be thought of as fossilized sunshine. More accurately, coal is fossilized organic plant material that grew in swamp-like conditions during prehistoric periods (Image 3.4). Coal is a black or brown sedimentary rock, composed primarily of carbon and related hydrocarbon molecules, derived from peat. Not only do coal deposits serve as a key energy resource, but also the study of fossilized plant matter in coal deposits has helped scientists reconstruct paleoclimates and ecosystems from periods of major coal formation.

Most commercially viable coal reserves come from several distinct periods of formation. No coal reserves predate approximately 420 million years ago, in the late Silurian period, because vascular plants had not yet evolved. Only when terrestrial vegetation

Image 3.4 These fossilized leaves in coal show the organic origins of the coal and help us reconstruct ancient ecosystems.
Bobo Deng / Imaginechina / Associated Press

colonized the formerly barren land and accumulated sufficient biomass to form peat did large coal deposits form. Three major geologic periods comprise the majority of commercial coal deposits. The first period dates to the **Carboniferous** (meaning coal-bearing) period, approximately 359 to 299 million years ago. Major coal deposits in the United States located west of the Appalachian Mountains date to this period and formed in massive deltaic wetlands in the region. The second major coal-forming episode dates to the Cretaceous period, approximately 146 to 65 million years ago. The Cretaceous had a hothouse climate, with no polar ice caps and high concentrations of atmospheric **carbon dioxide** (CO_2). The warmth and atmospheric conditions of a hothouse climate spur massive vegetation growth, which provides the basis for extensive coal deposits. Many of the deposits in Colorado and Wyoming date to this period. The last major coal formation dates to the Eocene epoch, approximately 56 to 35 million years ago, another hothouse climatic time. The extensive coal deposits along the Gulf of Mexico formed in this period, among many other global deposits.

The process of forming coal takes several million years (Image 3.5). The first step requires specific climatic and organic environments because coal forms primarily in wetland regions where extensive volumes of woody plants accumulate and are buried by sediments. Today, many areas could form future coal deposits, with peat bogs and swamps furthest along in the process. Peat is composed of organic plant materials that drop into standing water with low oxygen content. Because of the oxygen-poor environment, the organic matter does not decompose and instead accumulates. Peat bogs can be good at preserving organic matter for lengths of time that would be unthinkable in more oxygen-rich environments, with some European peat bogs holding artifacts such as well-preserved human mummies from long-past eras. The ability to preserve organic matter is a crucial part of the role peat bogs play in coal formation: over time, this preserved organic material is buried under layers of rock and sediment. The resultant heat and pressure of burial metamorphose the peat into various grades of coal, depending on the amount of time, heat, and pressure exerted on the organic deposits. The peat itself also contains sufficient organic matter that many people use it as an energy source. Although it is constantly being formed, peat is not considered a renewable resource because of its slow development.

Coal is classified based on its hardness and the relative amount of carbon within the rock. The harder and higher carbon concentration in the coal the more intense heat and pressure have been experienced by the rock layers, often meaning that such rock is also older. Coal is primarily broken down into four classifications, also known as ranks: lignite, subbituminous, bituminous, and anthracite. Higher rank coals burn at higher temperatures and burn cleaner than lower rank coals, which contain higher concentrations of other elements that produce pollutants other than CO_2 when burned. Although not technically coal, peat is also often included in classifications along with the various grades of coal, since it is the precoal material that leads to coal formation. The four grades of coal, plus peat, and their basic characteristics are as follows:

- **Peat**—a soft, wet organic material consisting of less than 50 percent carbon (Image 3.6). A traditional fuel source since ancient times, it is not in major use today.

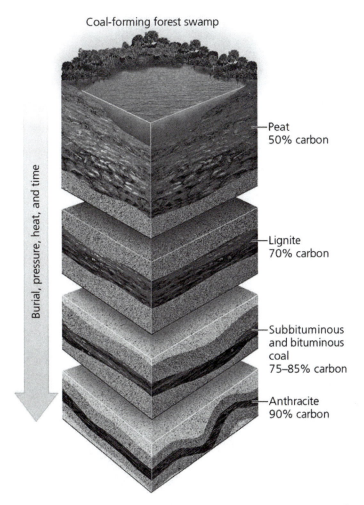

Coal-forming forest swamp

Burial, pressure, heat, and time

—Peat
50% carbon

—Lignite
70% carbon

—Subbituminous
and bituminous
coal
75–85% carbon

—Anthracite
90% carbon

Image 3.5 Organic material is compacted over time, creating peat. As the material is buried deeper, thus subjecting it to more temperature and pressure, it transforms into the different grades of coal, and the carbon content of each grows.

- **Lignite**—also known as brown coal, lignite ranges from about 50 percent to 70 percent carbon content (Image 3.7). It is a dirty fuel, often with both a high sulfur and a high water content, and is used primarily in electrical power generation.
- **Subbituminous coal**—this soft black coal ranges from about 70 to 80 percent carbon content, is lower in sulfur, and is commonly used in electric generation.
- **Bituminous coal**—this harder black coal ranges from about 80 to 90 percent carbon content and is used in electric generation and various industrial processes (Image 3.8). This grade is also the primary type used in gasification to

Image 3.6 Peat is a soft and wet organic material that can become coal with time, heat, and pressure changes.
Swapan Photography / Shutterstock

Image 3.7 Lignite, or brown coal, is the poorest class of coal.
Aleksandr Pobedimskiy / Shutterstock

make syngas, a fuel similar to natural gas. Coking coal is also generally produced from bituminous coal, which is used in steel making and other metallurgical processes.

- **Anthracite coal**—also known as hard coal, anthracite ranges from 90 to 97 percent carbon (Image 3.9). It burns slowly, although once ignited, it is hard to extinguish. It is relatively rare and expensive compared to the other grades of coal and is used primarily in residential and industrial heating processes, but not in electric generation.

Image 3.8 Bituminous, or soft coal, is a midgrade coal used often for electricity production.
Aleksandr Pobedimskiy / Shutterstock

Image 3.9 Anthracite, or hard coal, is the cleanest and slowest burning type of coal.
Lakeview Images / Shutterstock

Coal Mining

Peat was an important fuel source for many thousands of years, while coal began to be mined in more significant quantities around two thousand years ago in Britain. Coal seams can be found both near the surface and deep underground, leading to different methods of extraction based on the location of the resource and the available technology and labor. Until recently, coal mining worldwide was labor intensive and incredibly dangerous. Recent developments have improved and made surface coal extraction much safer, but many miners around the world, even in highly developed countries, continue to mine under dangerous conditions.

Until the second half of the twentieth century, most coal mining occurred in underground mines. **Underground mining** involves tunneling through layers of rock to reach a coal seam, which is then partially excavated and brought to the surface. This method is sometimes called deep mining because many coal mines extend more than 300 meters (984 feet) underground, with the deepest exceeding 1,500 meters (4,900 feet). Two principal methods are used to remove the coal: **pillar mining** and **longwall mining** (Image 3.10). In pillar mining, areas of coal are left intact to act as columns, while coal is removed between them to prevent the mine from caving in. In longwall mining, one face of the coal seam is removed at a time with a conveyor system in place along the mining face, with areas left intact to support the rock layers above, or the mine is allowed to collapse in the sections where the coal has been mined out (Image 3.11). Today, longwall mining and its relative, continuous mining, comprise the majority of underground mining in developed countries. Both modern techniques are highly mechanized, with large machines grinding the coal face into small pieces that can then be loaded onto vehicles and removed from the mine. This reduces the number of people needed underground at any given time and improves mine safety, although there are still significant numbers of people needed to work underground. These methods also produce far larger amounts of coal dust than premechanized extraction methods. Because of the immense

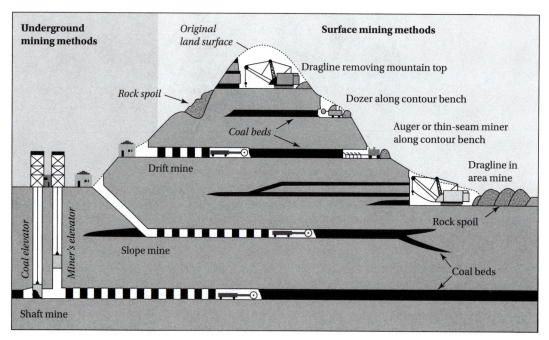

Image 3.10 Various methods can be used to mine coal, both above and underground.

National Research Council. 2007. Coal: Research and Development to Support National Energy Policy. Washington, DC: The National Academies Press. https://doi.org/10.17226/11977

Image 3.11 Machines like this are used for longwall mining in modern coal mines.
Shutterstock

weight of the rock above, even these methods of extraction do not entirely prevent cave-ins, although such incidents are less common now than in previous centuries.

The risk of cave-ins is only one of many hazards encountered in underground coal mines, which remain dangerous even in developed countries. In the past, and still today in some underdeveloped regions and "unofficial" mines, miners would be lowered in barrels or had to crawl through long passages to reach coal seams, and animals such as dogs and horses would be used to haul supplies in and coal of out mines. Insufficient or nonexistent safety precautions as well as the physically straining conditions have made mining an arduous and unpleasant profession (Image 3.12). Further hazards in subsurface mining include methane gas buildup, mine flooding, and insufficient ventilation. Poor ventilation can lead both to methane explosions and to asphyxiation of miners. Before modern sensor technology, miners would bring a caged canary with them into the mine; the bird would pass out and fall from its perch when methane levels got too high because it has higher sensitivity to methane's effects. The expression *a canary in a coal mine* as an omen of impending misfortune comes from this practice. Flooding has also trapped and drowned many miners when groundwater supplies leak into a mine, sometimes leading to catastrophic flows of water rushing through deep underground passages.

Underground mines are unpleasant, even in the most highly developed and regulated places. As coal is broken apart along its face, large amounts of dust are created, adding to the claustrophobic and hot environment. In regulated modern mines, miners wear respirators (Image 3.13); however, many miners around the world, including in developed countries, breathe in large amounts of coal dust, which leads to **black lung disease**, asthma, and other respiratory conditions.

With the invention of modern construction machinery and the development of TNT, a large amount of coal mining has moved from underground to **surface mining**, also

Image 3.12 This miner, after a long, hard day in the coal mine, is covered in coal dust. Exposure to this dust, especially in the lungs, leads to serious health problems, many of which are fatal.
Pittsburgh, Pennsylvania (vicinity). Montour no. 4 mine of the Pittsburgh Coal Company. Coal miner at end of the day's work / Photograph by John Collier / Library of Congress

Image 3.13 This coal miner in Ukraine wears a respirator to protect from dust particles in the lungs. DmyTo / Shutterstock

known as **strip mining** (Image 3.14). This type of mining can be used for coal seams up to about 60 meters (200 feet) under the surface. Large excavators and other digging machines, along with explosive charges to break up rock, are used to remove layers of soil and rock, known as overburden from above the coal (Image 3.15). The coal seam can then be scooped up and removed, and the overburden piles are regraded and replaced. Plants can then be introduced back over the site, which is then considered reclaimed. Despite these reclamation projects, the landscape is often altered forever. The greater impacts of surface mining on the environment contrast with its generally lower human impacts; surface mines are much safer, and exposure to coal dust is more limited than in

Surface mining

Top soil

Overburden

Shallow coal seam

Image 3.14 Strip mining, or surface mining, involves the removal first of overburden, the layers above the coal, followed by the removal of the coal seam below. In developed countries, governments often require reclamation of the land after mining is finished.
Courtesy of the U.S. Energy Information Administration

Image 3.15 Wyoming is home to the largest coal strip mines in the United States.
View into the Eagle Butte coal mine in Gillette, in Wyoming's Powder River Basin / Photograph by Carol Highsmith / Library of Congress

underground mines. Neither type of mining is low impact, and it is debatable whether the human or the environmental impacts are greater on any particular community.

One form of surface mining, **mountaintop removal**, has gained significant notoriety and evoked the most controversy in recent years, especially in the United States. Rather than removing coal seams located under relatively flat areas or hillsides, this method involves the demolition of an entire mountain summit, removing the rock into

Image 3.16 Most coal produced in the United States is shipped via rail from the mine to its site of consumption, such as a coal-fired power plant.

Konjushenko Vladimir / Shutterstock

the neighboring valley and essentially flattening the top of the mountain down to the coal layer. This practice has been adopted widely in the Appalachian Mountains.

In all coal mines, after the coal is removed, initial processing takes place on-site. This processing includes the removal of rocks, dirt, and other noncoal materials; the breaking or grinding down of the coal into easily transportable pieces; and occasionally processing to remove excess sulfur and ash. The coal is then loaded for transportation onward by rail, road, ship, or pipeline for final use elsewhere. For pipeline transit, coal is ground down and mixed with water into **slurry**, which can be piped. According to the Energy Information Administration, 72 percent of American coal is shipped via rail for at least a portion of its transit (Image 3.16).

Coal Combustion and Electrical Generation

After mining and transit, coal arrives at its destination for a variety of energy uses; in developed countries, these uses are primarily in the industrial and electrical generation sectors. In some places, coal is still used for home and commercial heating, although that usage has fallen off dramatically in the developed world. Coal also once served as the world's principal transportation fuel, powering trains and steam ships; now such use is limited to novelties such as historic train rides (Image 3.17).

Today, electrical power generation is by far the greatest consumer of coal. Large coal-fired power plants generate enormous amounts of electricity, with some of the biggest in the United States generating more than major hydroelectric dams, with capacities exceeding 3,000 MW (Images 3.18 and 3.19). Such capacities require vast amounts of

Image 3.17 Few modern railroads use coal-fired steam engines today; however, such engines once served as a primary means of intercity transportation.
Anya Ivanova / Shutterstock

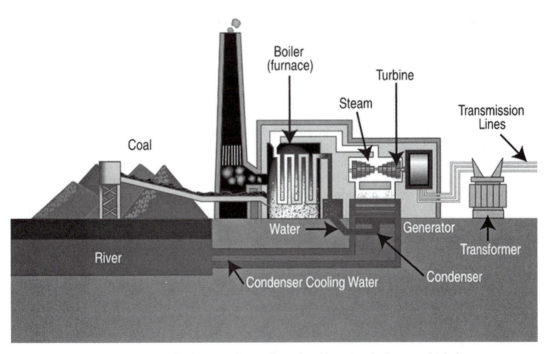

Image 3.18 Coal-fired power plants take coal and burn it to boil water, which then converts to steam, which turns a turbine. The turbine then turns an electrical generator and produces electricity, which can then be distributed to consumers.
Tennessee Valley Authority / Wikimedia Commons / CC-BY-SA-3.0

Image 3.19 This power plant in West Virginia burns coal to produce electricity.

coal, often many tons each minute to feed the boilers, and hundreds of rail cars a day during peak demand periods.

Coal power plants use massive quantities of coal each day to generate electricity. The coal is loaded onto conveyors, pulverized into a quick-burning fine dust, and then injected into a furnace with a boiler. The steam produced generates electricity with rotational turbines in the same manner as other steam-driven power plants. The flue gas produced by the burning coal is then sent up the smokestack or through **scrubbers** prior to venting. This gas contains many chemical compounds in addition to CO_2 and water vapor, some of which are removed in advanced facilities that employ scrubbers. Coal ash also goes up the stack unless filtered out, and more ash is removed in solid form from the combustion chamber. Ash consists of not only carbon, but also a slew of toxic substances including heavy metals such as mercury, which must be properly disposed of.

Coal is also a primary source of energy used in the manufacturing sector. Much of the coal for industry is used to generate heat, either to produce steam or in some cases to directly melt industrial products, especially for metallurgical purposes. Large inputs of coal have been used in steel making, which is still largely reliant on the coal industry in many parts of the world. Not only is coal used in the heating of metal for steel, but also some powdered coal is mixed in as an additive to melted iron in the making of steel (Image 3.20).

In the past, coal was also burned in significant quantities for use in steam engines. James Watt's improved steam engine used a boiler to create pressure, which was used to compress and expand a piston, turning a flywheel and creating power for transportation or industrial use. These engines, initially used in the early Industrial Revolution, powered machines in factories by turning rotors to drive belts and gears (Image 3.21). Soon thereafter, they were adapted to power the first locomotives, ushering in the era of the railroad. Using the same technology, steamships similarly increased the speed and reliability of maritime commerce and transit. The steam engine opened the American West to massive settlement expansion and the seas to the first era of

Image 3.20 Coal remains an important energy resource in the production of steel, both to provide heat to melt iron and as a feedstock to add carbon to iron and produce steel.

Kaband / Shutterstock

Image 3.21 In early factories, steam engines drove cranks and belts that then turned gears in machines, as seen with this belt-powered lathe.

Hans Christiansson / Shutterstock

globalization from the mid-1800s to 1914. The largest steamships, such as the RMS *Titanic*, burned up to 600 tons of coal a day while ferrying people and goods across the oceans (Image 3.22). Today, small-scale steam power is rarely used for transportation, with most active steam engines powering scenic rail and steamship tourist attractions rather than working trains and ships. For the most part, coal is no longer used as a

Image 3.22 The USS Pacific, a mail carrier ship, is one example of a steamship powered with coal, which dominated overseas shipping prior to diesel fuel.

U.S. mail steam ship Pacific: Collins line, builders, hull by Brown & Bell N.Y. engines by Allaire Works N.Y. / N. Currier / Library of Congress

transportation fuel , having been largely eclipsed by internal combustion engines over the second half of the twentieth century.

Worldwide Coal Production and Consumption

Since 2000, global coal production and consumption have increased every year until 2015. Despite this overall increase, the global pattern of coal production and consumption has varied. The fuel has lost both market share and overall production and consumption in developed countries in recent years, while growth in the developing world has outpaced the slowdown among developed countries. In developed countries, natural gas, renewables, and increasing energy efficiency have all led to decreases in coal usage. This drop in use in the developed world has been so marked that consumption is now lower than in the year 2000, although overall global coal production and consumption have risen significantly. By the end of 2012, developed countries only accounted for about 26 percent of total world coal consumption.

Coal is widely distributed around the world in sedimentary rock formations. However, certain locations possess much larger shares of these reserves than others. At current levels of consumption, the economically viable global reserves of coal will last another 130 years or longer. These proven reserves of coal are located overwhelmingly in countries that already produce large amounts of coal. According to the International Energy Agency, 75 percent of global coal reserves are concentrated in just four regions: North America (25.1 percent), former communist countries of eastern Europe and Central

Asia, also known as the transition countries (22.1 percent), China (18.5 percent), and Asia, excluding China (11.2 percent). Over 60 percent of these reserves exist in just four countries—the United States, Russia, China, and India—which together account for over 40 percent of the world's population.

Since 1984, China has been the world's leading producer of coal, despite smaller reserves than the United States and the former Soviet bloc (Table 3.1). Over half of all coal mined in the developed world comes from the United States and Canada, with several European countries also producing significant amounts.

Despite coal's widespread deposits and low price, its exploitation remains concentrated in a few key places. The top five countries that account for over 75 percent of all coal use globally are China, the United States, India, Russia, and Japan (Table 3.2). When the top ten are aggregated, the total rises to over 85 percent of global consumption. Most of the global growth is driven by China, which from the standpoint of calories

Table 3.1 Top Ten World Coal Producers for 2012 (in thousand metric tons)

China	3,549.1
United States	934.9
India	595.0
Indonesia	442.8
Australia	420.7
Russia	353.9
South Africa	259.3
Germany	197.0
Poland	144.1
Kazakhstan	126.0
World	7,830.8

Source: International Energy Agency, 2013.

Table 3.2 Top Ten World Coal Consumers for 2012 (in thousand metric tons)

China	2,794.8
United States	608.4
India	492.9
Russia	189.5
Japan	161.7
South Africa	140.2
South Korea	111.2
Germany	108.8
Poland	78.0
Australia	70.3
World	5,529.6

Source: International Energy Agency, 2013.

Image 3.23 Strip mining has overtaken underground mining as the leading means for coal mining in the United States.

Nneirda / Shutterstock

burned (rather than gross tonnage) crossed over 50 percent of global coal consumption in 2012. This was the first time since international energy records have been kept in which one country accounted for over 50 percent of all the energy from burning coal globally. India, another rapidly growing economy with rising energy needs, has also aimed to use coal as the primary growth fuel and may overtake the United States as the second largest consumer if U.S. use of coal continues to decline.

Global growth in coal consumption is being driven overwhelmingly by rising consumption of steam coal—coal used to produce steam in power plants that generate electricity. Despite rising demand for steel, metallurgy is not a major driver because coal for steel production is needed less and less in developed countries, where new technologies including heavy fuel oil injection and electric arc furnaces are becoming more common. There is some growth in the use of metallurgical coal in developing countries and the former communist countries; however, new technologies make this use of coal less and less important in the overall global coal market.

The global coal trade also continues to grow, along with rising consumption. In 2012 alone, global trade in coal rose 9.7 percent to reach record levels. Despite this, over 65 percent of all coal consumed globally is used in the country in which it was mined. A group of countries that are smaller consumers have profited greatly from the increase in coal consumption, becoming major international traders. Indonesia is now the world's largest coal exporter, followed by Australia, Russia, and the United States (Table 3.3). Despite massive domestic production, China dominates the global import market, purchasing enormous amounts of coal from other countries in the Pacific region (Table 3.4). In general, countries in east and south Asia are the most reliant on coal imports, with Japan, India, South Korea, and Taiwan being the next four largest importers. Japan has had to increase coal imports dramatically since

Table 3.3 Top Five World Coal Exporters for 2012 (in thousand metric tons)

Indonesia	382.6
Australia	301.5
Russia	134.2
United States	114.1
Colombia	82.2
World	1,255.3

Source: International Energy Agency, 2013.

Table 3.4 Top Five World Coal Importers for 2012 (in thousand metric tons)

China	288.8
Japan	183.8
India	159.6
South Korea	125.5
Taiwan	64.5
World	1,276.0

Source: International Energy Agency, 2013.

the tsunami of 2011, after which Japan's extensive nuclear power industry was suspended. Much of the spike in developed world coal use can be attributed to Japan's growth in coal consumption alone.

There is a major discrepancy between export and import figures for the global coal trade, likely attributable to poor record keeping. Global import figures are reported as significantly larger than export figures, with a 21,000 metric ton difference. Because these numbers are recorded and reported differently among different countries, companies, and international entities, all production statistics must be viewed as estimates.

Coal is not a completely fungible commodity; it does not command a standard price worldwide because grade and location determine its selling price. Coal prices are hugely influenced by the grade of coal (anthracite being most expensive and lignite cheapest) and are determined first by the mine itself, where factors such as labor, depth of the coal seam, and type of mine (strip versus underground) all contribute to the overall cost. Low sulfur content also drives up coal prices because it is costly to filter out sulfur from flue gas, and many countries require coal plants to emit low levels of sulfur dioxide. Because of the low cost of coal compared to the costs associated with transportation, coal prices for final consumers can vary dramatically from the spot prices at mines. Coal that is transported by ship incurs the smallest additional transportation cost, followed by rail, and coal that must be trucked is the most expensive. Because of these factors, it is hard to discuss any sort of global average price for coal, unlike other energy resources such as oil or uranium.

VIGNETTE
The Human Costs of Coal Mining in China

China is the world's largest producer and consumer of coal, and the fuel has played a huge role in the industrial and economic development of the country (Image 3.24). Over three-quarters of China's electricity is generated in coal-fired power plants, most of China's steel and metallurgical production is powered by coal, and coal is even used as the primary fuel for cooking and heating in many rural areas. Coal is an indispensable resource for the Chinese economy, but the country's dependence on the fuel has resulted in serious environmental and human costs.

Even before considering the effects of burning coal, the coal-mining industry in China has had a poor safety record. While many large mines account for much of the coal production, thousands of small unregulated (and illegal) mines operate with little regard for safety procedures and labor standards. Deaths in mines peaked in 2002 at almost seven thousand and have been decreasing as the government intervenes to close illegal mines and update safety procedures. The death rate per million tons of coal has dropped over 85 percent in the past fifteen years, as production continues to rise but deaths decline steadily.

The most visible impact of China's huge appetite for coal is the extreme level of pollution in many cities across the country. Smog, sulfur, mercury, and many other contaminants are released into the atmosphere through the large-scale burning of coal, making China home to some of the most heavily polluted cities on earth (Image 3.25). The particulates released by coal burning lead to major increases in pulmonary and respiratory diseases and may contribute to almost one million premature deaths per year by some estimates.

Smog levels sometimes reach such overpowering levels in some major cities that the government orders most citizens to stay inside and forces many factories to temporarily close. Deaths from disease and accidents are a major cost to the Chinese economy, costing billions of dollars per year in lost productivity. Even without considering coal's major role in China's status as the biggest emitter of **greenhouse gases**, the human, environmental, and economic costs of coal use are serious. Because coal has had such adverse effects, the Chinese government has committed to developing major alternative sources of energy, but coal use is expected to continue to rise until at least 2020, and these serious impacts will continue to harm the country and its citizens in the interim.

Image 3.24 China leads the world in coal mining and consumption.
Zhou Jianping / ICHPL Imaginechina / Associated Press

Image 3.25 Chinese cities exhibit the high costs of massive coal consumption, with smog and growing numbers of deaths and illnesses from related respiratory issues.
Jiang Jianhua / Imaginechina / Associated Press

Coal Production and Consumption in the United States

The United States remains the world's number two coal producer and consumer, by far the largest of the developed countries. The United States also holds the world's greatest coal reserves, with enough to last about 225 years at current rates of consumption (Image 3.26). Coal reserves in the United States spread across several regions. For most of U.S. history, the bulk of coal was sourced from mines in the Appalachian Mountains of the eastern United States. Recently, the majority has come from the western United States, especially from strip mines in Wyoming, primarily in the **Powder River basin**. In 2012, Wyoming was followed by West Virginia, Kentucky, Pennsylvania, and Texas as the top producing states.

The U.S. Department of Energy breaks the United States into three regions: Appalachian, Interior, and Western. Over half the coal produced in the United States comes from the Western region, primarily the Rocky Mountain province. Nine of the top ten producing coal mines in the United States are in Wyoming, with the Black Thunder and Antelope Rochelle mines producing more coal yearly than the entire state of West Virginia, America's second largest producing state. These strip mines are some of the largest coal mines in the world. Another third of American coal comes from the Appalachian region, where underground and mountaintop mining dominate production. The Interior region's production is more diffuse, with Texas accounting for one-third of the region's coal and smaller surface mines dominating production.

The United States is the global leader in coal reserves, with many experts calling it the Saudi Arabia of coal (Image 3.27). Only a small fraction of the coal located under U.S. soil lies in currently active mines. Beyond actively exploited reserves, geologists

Image 3.26 Coal reserves in North America by type and share of the world total.

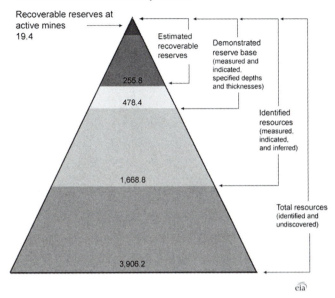

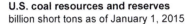

U.S. coal resources and reserves
billion short tons as of January 1, 2015

Recoverable reserves at
active mines
19.4

Estimated
recoverable
reserves

Demonstrated
reserve base
(measured and
indicated,
specified depths
and thicknesses)

255.8

478.4

Identified
resources
(measured,
indicated,
and inferred)

1,668.8

Total resources
(identified and
undiscovered)

3,906.2

eia

Image 3.27 The total amount of coal potentially under American soil far outpaces our ability
to mine it.
Courtesy of the U.S. Energy Information Administration

have calculated the estimated recoverable reserves, coal deposits that are documented
to exist but that have not yet been mined. Based on geologic information, much larger
estimated reserves can be assumed to exist, albeit with a somewhat lower degree of
certainty. Scientists estimate over 4 trillion metric tons of coal under U.S. soil when
including proven, estimated, and likely reserves, although it is highly unlikely that
most of these potential resources will ever be exploited. Of the U.S. reserves, 53.1 percent are bituminous coal deposits, 36.5 percent are subbituminous coal, 8.8 percent
are lignite, and only 1.6 percent are anthracite coal, located almost entirely in eastern
Pennsylvania.

Over time, U.S. coal production has shifted dramatically from the underground
mines of the east to the strip mines of the west. Eastern coal production has fluctuated
within a well-defined range for many decades. However, coal production in the western United States has skyrocketed, with Wyoming leading the way (Image 3.28). This
shift has occurred primarily because of the ease and lower cost of surface mining in the
western United States, as well as the lower sulfur content (Image 3.29). By the 1970s,
more coal was mined in strip mines than underground as a direct result of the boom in
western coal, culminating in the overtaking of the east in the late 1990s. This boom has
also caused the poorer, subbituminous coals of Wyoming to replace bituminous coal

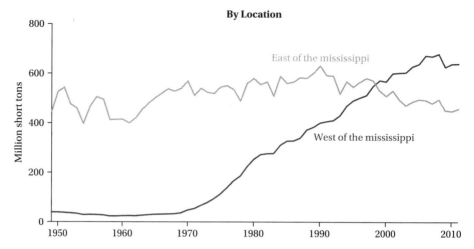

Image 3.28 Since the late 1990s, most coal mined in the United States has been in the western states.

Courtesy of the U.S. Energy Information Administration

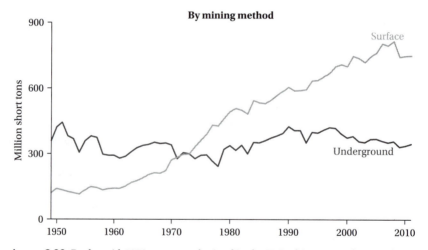

Image 3.29 By the mid-1970s, most coal mined in the United States was from surface mines, rather than underground.

Courtesy of the U.S. Energy Information Administration

as the primary grade of coal mined in the United States because of their higher carbon emissions and lower heat content (Image 3.30).

The rise of western U.S. coal has dramatically reshaped the American coal industry. Because of the mechanization of strip mining and its relative technical simplicity, production is an average of seven and a half times more efficient per man-hour than underground coal mining in the eastern United States.

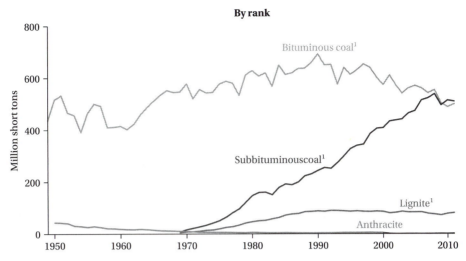

Image 3.30 The rise of coal mining in Wyoming has led to softer subbituminous coal becoming the primary grade of coal mined in the United States.

Courtesy of the U.S. Energy Information Administration

Most western coal is mined on lands owned by the U.S. government. In early 2016, the Department of the Interior announced a moratorium on new coal leases on federal land, citing local pollution impacts and future emissions from mined coal on the global climate. As a result, the boom in western coal may slow down, although current leases are estimated to remain productive for at least another twenty-five years. The moratorium has proven controversial, highlighting the tension between economic development and conservation.

In terms of domestic consumption, coal is overwhelmingly used for the generation of electricity in the United States. Approximately 93 percent of all coal burned in the United States is used for electric power generation, accounting for 37 percent of all annual electricity generation (equal to about 3.4 metric tons of coal per capita). The remaining coal is used by industry to produce both heat and coal byproducts. Certain component chemicals within coal such as ethylene are used for making plastics, among other uses. The steel industry, paper manufacturing, and concrete production also consume large amounts of coal in the United States.

Unlike at the global scale, it is possible to estimate coal costs within a national market, with good data on generalized coal prices in the United States. As noted in the global summary, different grades of coal, the mining techniques required, and transportation costs are all major factors in coal prices. According to the U.S. Department of Energy, in 2011 the average prices for coal at the mine site were $20.69 per metric ton for lignite, $15.51 per metric ton for subbituminous coal (because of the low production costs at massive Wyoming strip mines), $75.51 per metric ton for bituminous coal, and $83.44 per metric ton for anthracite. Western coal was generally cheapest, while Appalachian coal was most expensive because of labor inputs and the costs of mining underground.

Transportation costs can add significantly to the price of coal at final consumption points. On average, end users in the United States paid $5.82 per metric ton for transportation in 2011, reflecting 11 percent of the total purchase price. For coal-fired plants located far from the source, such as those in the eastern United States buying western coal, the cost of transportation can be as high as the cost of the coal itself.

Coal used for purposes other than steam production commands a far higher price on the market. Metallurgical coal (known as coking coal), used for steel manufacturing, cost an average of $203.31 per metric ton in 2011, almost four times the average price paid for coal in the United States that same year. Coking coal must undergo purification processes, sourced only from bituminous and anthracite types, making it more expensive.

American coal plays an increasingly important role on the global market. The United States is a minor coal importer, with imports consisting of about 1 percent of total coal burned in the country, primarily by power plants on the East Coast that are able to acquire cheaper coal from sources overseas. Almost 75 percent of imported coal comes from Colombia. The United States plays a far larger role as a coal exporter, currently the fourth largest in the world, with 114,100 metric tons of coal exported in 2012. During the 2000–2010 decade, the United States exported about 5 percent of its annual production, but in 2011 that number doubled, and U.S. coal exports continue to rise at a fast pace, driven by rising international demand (Image 3.31). American coal is mostly exported to Europe, with the Netherlands as the leading consumer (much of this coal is shipped to other countries in Europe from the massive port at Rotterdam), followed by Brazil, the United Kingdom, Japan, and Canada.

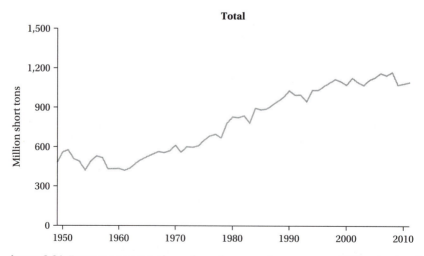

Image 3.31 European countries have always been a major consumer of U.S.-mined coal.
Courtesy of the U.S. Energy Information Administration

VIGNETTE

Mountaintop Removal Mining in the Appalachians

Traditionally, coal mining in the Appalachian Mountains has been underground, while mining in the American West has utilized strip removal and open pit techniques because of flatter terrain and less population density. Since the 1970s, however, the growth of surface mining in the more mountainous regions of West Virginia and Kentucky has resulted in what is known as **mountaintop removal mining**. Because the coal sources of the Appalachians are generally too mountainous to simply dig large pits, as most strip mining does, mountaintop removal physically removes the mountaintops of overburden above coal seams, allowing the coal to be removed from the surface (Image 3.32). The massive amounts of waste rock from the process are then disposed of in nearby valleys, known as valley fills.

Mountaintop removal is extremely controversial because many serious environmental impacts cannot be easily mitigated. The process destroys the original landscape, and even after the land is reclaimed after mining is complete, the natural contours are never returned to their premining state (Image 3.33). Waste rock, often contaminated by mine tailings containing toxic compounds, is dumped in valley fills, where it blocks rivers and streams and can leach into local water sources and ecosystems. The pollution of local water systems and the release of particulates into the air create major health problems for nearby populations in addition to the damage done to the landscape and natural systems.

Despite the environmental and human impacts of mountaintop removal, the process is favored by mining companies for several reasons. When overburden is removed at such a scale, larger machinery can be used to process coal much more efficiently, which lowers the costs of production and allows fewer miners to accomplish the same task. Because the miners are working above ground, there is no risk of collapse, making the process much safer than underground mining for the workers on-site. In the Appalachian states where coal mining makes up such a major part of the local economy, these benefits have often been touted as significant enough to justify mountaintop removal. As coal use in the United States continues to decline, the economic arguments in favor of mountaintop removal will likely be less persuasive in allowing the practice to continue, but for now, the practice remains a major part of the coal extraction industry in the Appalachians.

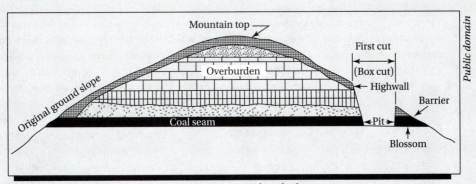

Mountaintop removal method.

Image 3.32 Mountaintop mining is a means of strip mining where an entire hilltop is removed into the valley below to remove layers of coal.
Coal River Folklife Project collection (AFC 1999/008), American Folklife Center, Library of Congress

continued

continued

Image 3.33 This site in West Virginia demonstrates the damage to the landscape caused by mountaintop mining.
David Stephenson / Shutterstock

Impacts of Coal Use

When one thinks of coal, air pollution immediately comes to mind. Coal produces many pollutants and causes numerous serious impacts during all stages of its production and consumption. Coal contains sulfur, chlorine, phosphorus, mercury, and other trace elements that pollute our air and water. When burned, it also produces ash and enormous quantities of CO_2. Of the major energy resources used today, coal is one of the most problematic for the environment. The coal industry often touts "clean coal" as a solution to this problem, although this is primarily a political tactic and advertising slogan to promote more coal usage. Clean coal is a broad term that coal advocates use to describe carbon capture and sequestration, scrubbing technologies, and more efficient combustion technologies. However, this description is highly misleading because even clean coal produces numerous serious environmental impacts during mining and combustion. Technologies that would make the combustion process less damaging are being tested, but these remain expensive and rare. Even if coal could be burned and all emissions sequestered, the major impacts of mining would continue, making completely clean coal virtually impossible. Despite its impacts, coal is cheap and ubiquitous and will therefore likely remain a major source of energy for some time.

Coal mining is environmentally damaging. Strip-mining techniques, especially mountaintop removal, have more pronounced localized impacts than underground mining. Strip mines can cover many square miles, and the removal of thick coal seams and massive amounts of rock necessarily entails major disturbances to local vegetation, water resources, and topography. Even after reclamation, there is no way to return the site to its premined conditions. This impacts both the environment and the aesthetics of the location. With mountaintop removal, these effects are even more pronounced because the mountaintop is dumped into neighboring valleys, permanently altering the region's topography and landscape while severely disrupting the water drainage system (Image 3.34). Water that runs through strip mines and the backfill from them picks up harmful acidic compounds, which are then carried downstream.

Although they are less invasive than strip mines, underground mines also create major impacts. Trapped methane deposits must be vented from the mines to avert explosions, and methane emissions are responsible for significant climate change effects because methane is a far more potent greenhouse gas than CO_2 by volume. In the United States, 7 percent of methane emissions and over 1 percent of total greenhouse gas emissions come from venting underground mines. Underground mines also contribute to acidic runoff problems, like surface mines, because water leaks down into the mines and can then escape into the area's water resources (Image 3.35). Mine tailings also further pollute the water.

Coal produces numerous harmful air pollutants (Image 3.36). Beyond CO_2, the effects of which are covered in Chapter 15, coal burning releases sulfur, nitrogen oxides, lead, arsenic, selenium, uranium, and mercury. Sulfur combines with oxygen to form sulfur dioxide, which is the principal cause of acid rain. Nitrogen oxides contribute to smog creation, and mercury and the other heavy metals fall back to the earth and enter the water system. Soot and ash released during combustion are breathed by populations downwind from the burning coal, contributing to respiratory problems.

Pollutants from coal mines leach directly into the water supply, and heavy metals, nitrogen oxides, and sulfur dioxide from rain are major components of coal-based water pollution. Additionally, coal ash slurry has broken out of containment in some cases, causing massive localized water pollution with major toxins (Image 3.37). Sulfuric acid–contaminated water, often laced with lead and mercury, leaches out of coal mines, polluting the watershed. This highly acidic water greatly harms plant life and fish

Image 3.34 Mine tailings show the long-term costs of mining, wherein toxins leach from mines into local water supplies.

Ian Woolcock / Shutterstock

Image 3.35 Acid runoff is one problem impacting water at mining sites. Courtesy of Phil Hill / Science Source.

Phil Hill/Science Source

Image 3.36 The burning of coal is a major contributor to air pollution.

Stripped Pixel / Shutterstock

in the contaminated bodies. Furthermore, the heavy metals released into the contaminated bodies **bioaccumulate** in higher order aquatic life. These metals impede mental development in humans and animals and can be toxic to life. Acid rain destroys forests and kills aquatic plants, harming wetland ecosystems.

The burning of coal is the world's single largest contributor of anthropogenic CO_2 emissions. In 2010, coal combustion released 13.1 billion metric tons of CO_2 into the atmosphere. In the United States, although coal only generates a little over one-third of total electricity, it contributes almost three-quarters of all greenhouse gas emissions released

Image 3.37 Coal ash can be released accidentally, as happened in 2008 in Harriman, Tennessee, where a retention wall failed, releasing coal ash into the river.
Wade Payne / Associated Press

through electrical generation. Coal has a higher carbon-to-hydrogen ratio than other fossil fuels, which causes it to emit more CO_2 than the equivalent mass of oil or natural gas.

For coal miners, especially those working in cramped underground mines, many personal risks lead to shorter life spans, especially in places where regulations are lax or nonexistent. Cave-ins and underground explosions trap or instantaneously kill thousands of coal miners a year around the world. China is most notorious for these accidents, along with Russia and Ukraine, but such accidents even occur regularly in the United States. A recent major accident in the United States, the Upper Big Branch Mine accident in 2010, led to the deaths of twenty-nine miners when coal dust exploded in the mine. The world's worst coal mine accident occurred in China in 1942, with over fifteen hundred fatalities.

Other than underground accidents, a far more common problem comes from coal worker's pneumoconiosis, better known as black lung disease (Image 3.38). This condition is caused by long-term exposure to coal dust and is especially prevalent in miners who work underground. The disease is caused by coal particles adhering to the lungs permanently, which impedes the body's ability to absorb oxygen over time and can lead to death. The Mine Health and Safety Act of 1969 led to U.S. federal standards for coal miners, including standards for the amounts of coal dust in a mine and for the provision of personal respirators. Such protections are not available to miners in less developed parts of the world, and even in highly developed countries, including the United States, 5–10 percent of miners are likely to develop the disease.

One of the biggest challenges regarding the impacts of coal use comes from the lack of a tie between the end user of coal and the impacts. Most of the costs of coal use are indirect, in that those impacted are separate from the big consumers. Major power utilities do not directly deal with mine accidents, the impacts of air and water pollution, and other coal impacts. Because the major coal producers and consumers are not responsible for the financial and social costs that their resource use drives, they have little

Image 3.38 Coal dust impacts coal miners and can lead to black lung disease.

DmyTo / Shutterstock

incentive to opt for a cleaner approach. Regulations such as carbon taxes can transfer these costs to the entities that cause them, but until such rules become widespread, coal's low price for electric generation is likely to ensure its dominance.

Mitigating the Impacts of Coal Use

Although there is no such thing as clean coal, attempts are being made to make coal much cleaner than it has been. Because of the massive energy needs of modern society, implementing cleaner coal technologies and researching ways to burn coal with fewer impacts would be a practical approach to maximizing resource usage while minimizing climatic and health impacts. There is no conceivable approach in the short to medium term that excludes coal from global energy use.

In developed countries, scrubbers are now widely required to reduce the air pollution impacts that occur where coal is burned. Scrubbers for coal plants typically spray water and various chemicals into the exhaust after combustion, trapping much of the particulate matter as well as a portion of gases like sulfur dioxide and nitrogen oxides, but not CO_2 (Image 3.39). These systems are expensive to retrofit into existing plants, and although they are cheaper to install on new facilities, they still add significantly to costs. Additionally, wet scrubbers create large volumes of liquids full of ash and pollutants that must then be stored or treated, and liquid storage must be secured to keep

Image 3.39 Scrubbers help capture particulates, sulfur dioxide, and other pollutants in flue gas from a coal-fired plant, but they do not capture carbon dioxide.
Ph.Wittaya / Shutterstock

it from entering the water supply. To comply with the Clean Air and Water Acts, new coal-burning facilities must install scrubbers in the United States.

To make significant progress in limiting coal's impact on the global climate, significantly less coal must be used, and the greenhouse gases from coal combustion will need to be kept out of the atmosphere. **Carbon capture and sequestration** offers a solution to the second need; it involves separating CO_2 from the flue gas, condensing it into liquid form, and then either injecting it underground or piping it elsewhere (Image 3.40). However, carbon capture and sequestration on the scales needed for large power plants is only in the early testing phases. Carbon-capturing systems are also energy intensive, and a plant likely would have to consume from 25 to 40 percent more coal to produce the power needed to remove the CO_2 from its flue gas.

The other means to reduce carbon load while burning coal is to pretreat the coal through gasification. This conversion of solid coal into a gaseous state is done prior to burning the coal. This process allows for more efficient electrical generation in a two-cycle plant (discussed further in Chapter 5 on natural gas); however, the costs to build and run such a plant are significantly higher than that for a conventional coal-fired power plant.

An experimental form of coal-gas production involves the injection of heated oxygen and other gases directly into coal seams, which then bond with parts of the coal and come back to the surface well as gases that can be burned like natural gas in a power plant (see more in the coal-bed methane discussion in Chapter 5 on natural gas). This method has the additional advantage of avoiding many of the ecological and human impacts of coal mining.

Coal is a major part of the global energy system, and its major environmental and social impacts represent a significant challenge in achieving a clean and sustainable future. Most countries rely on coal for at least a portion of their energy supply, and consequently coal will

Image 3.40 Carbon-capture technologies aim to trap carbon dioxide released by combustion in power plants (left) and oil pumps (center) and inject it into deep strata where it will be isolated (right).
Nicolle R. Fuller/Science Source

continue to be used in the short to medium term. Technology advancements could reduce the impact of this widespread use, although transitioning away from coal to other energy sources is likely to be the long-term solution to coal's many problems. Lower natural gas prices in the United States have already pushed a shift away from coal, but without regulatory changes that shift the burden onto producers and end users, global coal use is likely to continue to rise for the foreseeable future. Given coal's ubiquity and disproportionate impacts, its use remains one of the greatest problems to overcome in the twenty-first century.

REVIEW QUESTIONS

1. Coal first emerged in the Roman period for what purposes? Are any of these still used today?
2. What are the four ranks of coal? Why do we not use them interchangeably and why is peat not one of them?
3. What are some of the hazards faced in underground coal mines? Can these ever be 100 percent protected against?
4. Why is coal primarily used only in electrical production and industry today? How did that differ one hundred years ago?
5. Why do China and the rest of east and south Asia account for most global coal imports? Is most coal sold internationally or domestically?
6. Where in the United States is coal still mined underground? Where do surface mines dominate?
7. Why is mountaintop removal one of the most controversial forms of coal mining?
8. In what ways do coal production and consumption contribute to water pollution?
9. What are some methods we can use to mitigate some of the impacts of coal on the environment?

Image 4.1 Vladimir Melnikov / Shutterstock

CHAPTER 4

Oil

Crude **oil** found uses by humans long before its rise as an energy source. Forty thousand years ago, early humans employed **bitumen**, a form of crude oil, on certain stone tools. In ancient Mesopotamia (modern Iraq), bitumen was harvested for use in building materials and to waterproof objects. It was also traded to other cultures, such as ancient Egypt, where it aided in mummification. Crude oil became a medicine in some cultures and was used for weapons in ancient Persia. By the Middle Ages, some cities in Iraq used tar, another crude oil derivative, to pave roads.

Although coal drove the Industrial Revolution, oil emerged as the world's principal source of energy in the twentieth century. Prior to the mid-nineteenth century, oil was not produced at a commercial scale because subsurface deposits are more technically challenging to exploit than was generally possible at the time. Surface deposits and oil seeps were well known and widely exploited. This changed in the late 1850s, when the first oil wells were drilled in Pennsylvania and Canada (Image 4.3). The advent of and rapid advances in drilling technology allowed oil to be economically extracted, paving the way for its growth as a major energy resource. At first, oil was primarily produced for use in kerosene lamps, replacing whale oil and candles for lighting. Later, it took off as the power source for internal combustion engines in lieu of steam engines, leading to the oil-based global economy we have today.

The oil boom was one of the most important events in the development of modern industry, foreshadowing the incredible power that oil would hold for the economy and geopolitics during the twentieth and into the twenty-first century. Early major discoveries in California and Texas turned those states into massive oil producers,

Image 4.2 This 1930s well in Iraq is producing bitumen, a heavy form of crude oil that has been used for many centuries.

Library of Congress/Science Photo Library

Image 4.3 Titusville, Pennsylvania, was one of the first places in the world where oil wells were drilled.

Photo by Mather / ca. 1900. / Library of Congress

kicking off decades of economic and demographic growth. The Kern River field in California, first drilled in 1899, has been continuously delivering oil to the present day, and more than a century later it remains one of the top ten producing fields in the United States. The Spindletop field in Texas, first produced in 1901 and etched into history for its gushing wells, made over 100,000 barrels per day during its peak, triggering the Texas oil boom that would transform the state into one known for energy production. Although the United States has since been surpassed by two other countries in oil output, these early discoveries made it the largest oil producer in the world and built the fortunes of companies like Standard Oil, run by the Rockefeller family (Image 4.5). The large reserves of oil in the United States played a pivotal role in the

Image 4.4 Oil can naturally seep to the surface, which allowed it to be collected in small amounts long before the development of oil-drilling technologies.

Sinclair Stammers/Science Source

Image 4.5 John D. Rockefeller was the most famous leader of Standard Oil, the dominant force in the early American oil industry.

Copyright by George G. Bain / c1910 / Library of Congress

two World Wars as well because Western Europe produced little oil, making American production essential in keeping the Allied war effort moving. Since then, oil has continued to play a significant role in geopolitics, casting the Middle East into a major economic region, bringing the cartel of the Organization of the Petroleum Exporting Countries (OPEC) to world prominence, and setting off conflicts worldwide. Unconventional technologies have since rejuvenated oil production in the United States, leading to a reduced dependence on imports and a rapidly fluctuating energy price market beginning in late 2014.

Oil has become the dominant energy resource of the modern era for many reasons, chief among which are its compactness and liquid form. Oil is energy dense, making it relatively cheap to transport for its comparative power. This characteristic also singles out oil as the best base for transportation fuel because it can be easily carried in sufficient quantity without weighing a vehicle down (Image 4.6). For reference, 1 kilogram of anthracite coal contains around 38,000 kJ of energy, whereas a kilogram of gasoline (about 0.75 liters or 0.2 gallons) contains around 44,000 kJ. This higher energy density, along with the ability to be transported by **pipelines** and tankers at high efficiencies, makes oil a highly versatile fuel source.

In addition to its high energy content, oil plays an essential role in the global economy with its refined products (Image 4.7). Although gasoline may be the most well-known derivative of oil refining, plastics, soaps, and even food products like bubble gum come from products refined from oil. Even after society moves on from using **petroleum** as the primary energy source for transportation, it will likely remain dependent on the extensive range of products made from oil that have pervaded almost every part of the modern world.

Image 4.6 The high energy density of oil makes it particularly useful for transportation, whether by land, by sea, or by air.

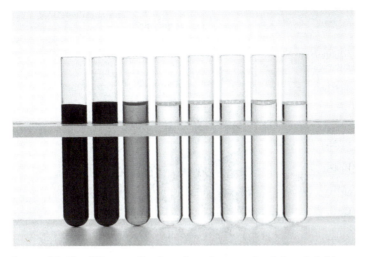

Image 4.7 The different refined products from crude oil, from left: bitumen; fuel oil; heavy, medium, and light lubricating oils; diesel fuel; jet fuel; and gasoline. Courtesy of Paul Rapson / Science Source.
Paul Rapson /Science Source

Origins and Geology of Petroleum

Like other fossil fuels, oil is formed over time, deep underground, from ancient organic material. Although popular culture represents oil as being derived from dinosaurs and other prehistoric creatures, **phytoplankton**, along with other microscopic oceanic organisms that accumulated on the seabed millions of years ago, accounts for the bulk of its composition. Assuming these creatures die and collect in sediments in a relatively oxygen-poor environment, they will eventually build a portion of a **stratum**, or geologic layer. As more strata form over this layer over thousands or millions of years, it will be compressed and heated by conditions deep below the surface. After time at high temperatures, the organic material converts into **kerogen**, the precursor to oil and natural gas. As temperature increases through burial, the basic structure of the organic hydrocarbon chains chemically changes, analogous to a cooking reaction (Image 4.8). Assuming the material is consistently heated at the right temperature through time, it will form oil source rock, if sufficiently rich in organic matter. Further pressure and higher temperatures will continue to transition the kerogen into other products, beginning with liquid oil and continuing to condensates of petroleum all the way to natural gas with no liquids at the hottest end of the spectrum. The conditions in which oil forms geologically explain why it is relatively rare; insufficient temperature will prevent the formation of oil, while too much temperature will destroy liquid oil and even convert natural gas into carbon dioxide (CO_2) in the presence of sulfur.

In addition to temperature and pressure, oil only occurs in specific types of rock in quantities that are economical to produce. Because large amounts of organic material must be deposited in layers to give sufficient material that can form oil, oil source

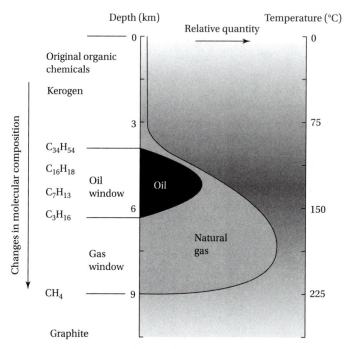

Depth (km)   Temperature (°C)

Relative quantity

Changes in molecular composition

Original organic chemicals

Kerogen

$C_{34}H_{54}$

$C_{16}H_{18}$

Oil window

C_7H_{13}

C_3H_{16}

Oil

Gas window

Natural gas

CH_4

Graphite

Image 4.8 Oil only forms under specific temperatures and pressures, as seen in the oil window. Information from University of California, San Diego

rocks are exclusively fine-grained **sedimentary** rocks known as mudstones and **shales** (Image 4.9), typically black to dark brown in color and rich in organic matter derived from algae. Oil can accumulate in porous rocks such as **sandstones**, limestones, and dolomites through migration from source rocks as heat and compaction squeeze oil and its precursors from the source rocks. These rocks, often referred to as oil and gas reservoir rocks, have several characteristics that allow oil and gas to be successfully and economically extracted. Sandstones and other sedimentary rocks can be porous, meaning that a significant part of their rock structure is made up of empty space within the rock. The porosity within these rocks that can hold reservoir oil are not caverns or pools; rather, they are void spaces between grains or other small spaces. These empty spaces would initially be filled with water following sedimentary rock deposition and burial. Porosity can also be created in rocks by mineral reactions with groundwater, which can remove rock material and thus create pore volume. Porous rocks can thus become saturated with oil and, under certain conditions, large volumes of this rock store quantities of conventional oil in the subsurface. These conventional oil fields can contain up to billions of barrels of recoverable petroleum.

Until recently, shales, mudstones, and other fine-grained sedimentary rocks were of little interest to the oil industry because they were too impermeable for commercial oil or gas production. Low permeability means that fluids cannot easily flow in

Image 4.9 Shale, a type of sedimentary rock, is denser than sandstone, and only with fracking processes has any large amount of oil been produced from shale.

Sigur / Shutterstock

or out of the rock because of the small size of the pore spaces and the connections between the pores. Consequently, shale oil, a type of unconventional oil, requires the use of newer extraction technologies such as **hydraulic fracturing**, sometimes referred to as *fracking*, to flow into wells at commercial rates and volumes. Some oil-saturated rocks contain oil that has been degraded into *heavy oil* or tar, usually by bacterial action. These types of unconventional accumulations require heating of the rock by steam injection or other means to lower the viscosity of the oil and allow it to flow through the pore space or, in the case of shallow **tar sands** such as those in western Canada, the deposits are strip mined and the oil extracted by heating the tar sands in a mixture with water.

Refining and Uses of Oil

Oil is used to create a wide variety of products, from fuels to plastics, all of which must be refined from **crude oil**, which is oil's natural state. Oil comes out of the ground in varying compositions because of the variations in organic compounds from which it formed, and it therefore varies in how much of each end product it can produce. To separate the wide variety of compounds within crude oil, it must be processed at an **oil refinery**, where these products can be separately extracted or modified by chemical reactions and sent to end users. Through the refining process, gasoline, diesel fuel, kerosene, butane, propane, heavy fuel oil, waxes, lubricants, and plastic precursors are all separated out from crude oil (Image 4.10).

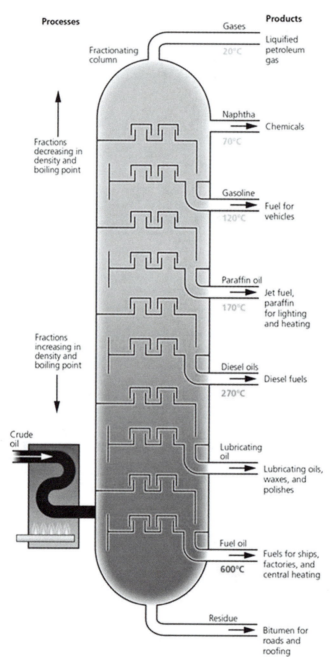

Processes

Fractionating
column

Fractions
decreasing in
density and
boiling point

Fractions
increasing in
density and
boiling point

Crude
oil

Products

Gases
20°C — Liquified
petroleum
gas

Naphtha
70°C — Chemicals

Gasoline
120°C — Fuel for
vehicles

Paraffin oil
170°C — Jet fuel,
paraffin
for lighting
and heating

Diesel oils
270°C — Diesel fuels

Lubricating
oil — Lubricating oils,
waxes, and
polishes

Fuel oil
600°C — Fuels for ships,
factories, and
central heating

Residue
— Bitumen for
roads and
roofing

Image 4.10 The refining process separates lighter liquids with lower boiling points at the top of the column from denser ones with higher boiling points at the bottom.

To begin the refining process, oil must first be heated, and then it can be subjected to **fractional distillation**, which allows its constituent parts to be separated. Distillation relies on the weights and densities of these different components to allow them to congregate in layers, much as oil floats on water. The gaseous components of oil such as butane and propane will rise to the top of the distiller, and heavy components will separate further down, leading all the way to materials as heavy as asphalt and tar at the bottom. The actual process is far more complicated than this and can involve reprocessing heavy products, filtering out contaminants such as sulfur, recombining fuels, and other applications. Consequently, oil refineries are large and expensive operations, and they come with many environmental concerns. Explosions at refineries are a major concern because of the volatile products they produce, and refineries are also a significant generator of air and water pollutants and greenhouse gases. Because of the cost of building new refineries, due to both their scale and ensuring their environmental compliance, no new plants have been built in the United States since 1976. Refining rates have stayed relatively constant since then because of renovations and expansions of existing plants; no new refineries likely will be built in the coming years.

The primary products of an oil refinery include a wide variety of fuels, along with other oil-derived products. In the United States, the average barrel of oil is primarily refined to produce transportation fuels, with gasoline and diesel accounting for the largest shares. The amounts refined by an average barrel and their uses are as follows (Image 4.11):

- **Gasoline** (42 percent of an average refined barrel of crude oil in the United States)—primarily used as a motor fuel in cars, light trucks, and other small machines. Gasoline is primarily refined crude oil, but also contains other additives placed in the fuel at the refinery. It is refined usually into three grades—regular, midgrade, and premium—which are differentiated by the **octane rating**, a measure of the pressure required to ignite the fuel. Higher octane fuels command a higher price for consumers.
- **Diesel** (27 percent of an average refined barrel of crude oil in the United States)— primarily used as a fuel in freight trucks, trains, buses, boats, and farm equipment, although also in some cars and light trucks. Diesel is also used as the fuel for many backup electrical generators used in hospitals, other large buildings, and even some private homes, along with some isolated communities. Diesel is denser than gasoline and requires a different engine design.
- **Heating oil** (2 percent of an average refined barrel of crude oil in the United States)—closely related to diesel fuel, heating oil often has a higher sulfur content. It is often used in the United States as a home-heating fuel, especially in the northeastern states.
- **Jet fuel** (9 percent of an average refined barrel of crude oil in the United States)— as its name suggests, this fuel powers jet aircraft. It is less dense than diesel, but denser than gasoline.

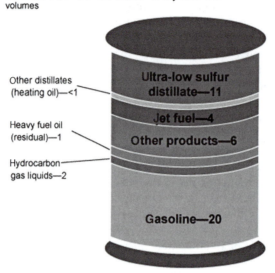

Petroleum products made from a barrel of crude oil, 2016
volumes

Other distillates (heating oil)—<1

Heavy fuel oil (residual)—1

Hydrocarbon gas liquids—2

Ultra-low sulfur distillate—11

Jet fuel—4

Other products—6

Gasoline—20

Note: A 42-gallon (U.S.) barrel of crude oil yields about 45 gallons of petroleum products because of refinery processing gain. The sum of the product amounts in the Image may not equal 45 because of independent rounding.

Image 4.11 A barrel of crude oil is refined into many different products. In the United States, gasoline is the most important product produced.
Courtesy of the U.S. Energy Information Administration

- **Liquefied petroleum gases** (4 percent of an average refined barrel of crude oil in the United States)—lightweight fuels such as propane, ethane, and butane are the primary components of liquefied petroleum gases. They can also be produced from natural gas. Liquefied petroleum gas is compressed and cooled into liquid form for easy transportation and storage, but it becomes gaseous when released from its pressurized container. Cooking fuel is among its many other uses. It is the lightest grade of crude oil product.
- **Heavy fuel oil** (less than 2 percent of an average refined barrel of crude oil in the United States)—this fuel is used primarily in industrial plants and large-scale oil-fired power plants (Image 4.12), as well as shipping, sometimes called bunker oil. Although not used widely in the United States, a large portion of electricity is generated with heavy fuel oil in some countries, such as Saudi Arabia. It is the densest fuel-grade substance in crude oil.

A variety of other products compose crude oil's remaining 13 percent, including basic industrial products such as tar and asphalt. Other feedstock chemicals are produced that are used in the manufacturing of plastics, food coloring, chewing gum, types of glue, and countless other consumer products.

Image 4.12 Some power plants use fuel oil or kerosene to boil water and run a turbine to produce electricity. This plant uses these products in addition to natural gas.
Felix Lipov / Shutterstock

Finding and Producing Petroleum

As world demand for oil has increased exponentially over the past century and a half, methods of finding and extracting it have evolved from simple operations into massively complicated and expensive undertakings by some of the largest corporations in the world. Finding oil requires a number of geological, economic, and logistical conditions to be met, making it an often-risky proposition, but one with potentially massive monetary and energy returns. In basic terms, a successful well will be part of a larger **reservoir** that is made up of one or more oil-bearing rock strata, which are known as the **producing formation**. By drilling into the producing formation, oil can be accessed and brought to the surface for transport and further refinement into fuels and other products. While this process is conceptually simple, a number of things make it difficult, expensive, and risky.

Early oil fields were technologically simple compared to modern petroleum extraction and relied on shallow reservoirs that could be tapped with basic drilling techniques. Before the scientific development of petroleum geology, fields would often be found from clues on the surface. Oil seeps or exposed strata with oily residues had often been known by locals for hundreds of years in many oil-producing regions (such as the La Brea tar pits in central Los Angeles [Image 4.13]), with many of these sources having been exploited long before the modern development of the oil industry. These surface clues generally indicate the presence of much larger quantities of oil below the surface, making them obvious targets for further exploration. Seeps led to the development of many of the first fields in the United States, from the first major oil well in Titusville, Pennsylvania, to the massive oil fields of East Texas and Kern County, California (Image 4.14).

Image 4.13 The La Brea tar pits in Los Angeles, California, are a famous example of an oil seep.
Larissa Pereira / Shutterstock

Image 4.14 Early famous images of oil production include gushers, such as this photo at the Spindletop oil field in Texas. High pressure in an oil reservoir would lead to crude oil gushing out of the oil derrick until it could be capped and then produced.
The Bernier Publ. Co., N.Y. / c1901. / Library of Congress

In the early, shallow fields, the first wells were drilled using **percussion drilling**, which used a hammerlike action and gravity to punch a drilling cable tool through rock into the target formation. Percussion drilling is inefficient, unreliable, and unable to reach deep, so early oil exploration was geographically limited. Once drilling reached the targeted formation, early wells relied entirely on high **reservoir pressure** to push the oil up the drilled hole to the surface on its own. These wells usually had high pressures, at least initially, leading to the famous **gushers** that many people associate with oil production, in which oil explosively shoots from the drilling rig into the air. Once reservoir pressure has been relieved, oil pumps continue to extract from the reservoir until production levels are too low to overcome the costs of continued extraction (Image 4.15).

Once the limitations of early percussion drilling techniques became clear, oil explorers developed the more modern **rotary drilling** systems that have continued to evolve until today. Rather than hammering through rock formations, rotary drills used a drill bit that rotates in place, grinding rock away (Image 4.16). By pumping fluids into the hole at the same time as drilling, the drill bit could be kept cool while simultaneously flushing drilled material back to the surface, greatly increasing the speed and efficiency of the drilling process. This method of drilling vastly increased the depth at which oil could be extracted, opening many more regions to oil exploration and production. In time, the science of oil exploration developed as well, adding many more tools for oil exploration. Rather than simply drilling near oil seeps, scientists could examine geological maps to look for particular rock formations and then refine their search with many different physical tools. **Seismic imaging**, which uses sound waves to penetrate the surface, allows explorers to see the physical structure of rock thousands of feet below the surface, while chemistry and geology allow scientists to determine how much organic content, temperature, and pressure to expect in any given area, thus giving an improved picture of where oil might be found.

After the types of source rock that produce petroleum are identified, areas that hold oil must be pinpointed before oil can be successfully extracted. Because oil is a liquid, it

Image 4.15 Once pressure drops in an oil reservoir, pumps can be used to continue to produce oil from the well.

PhotoStock10 / Shutterstock

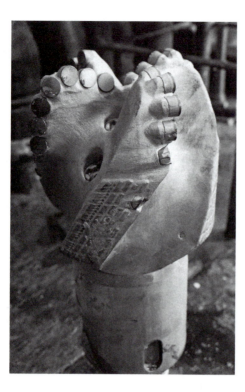

Image 4.16 Rotary drill bits allow for efficient drilling through thousands of feet of rock to reach an oil reservoir underground.

James Jones Jr / Shutterstock

can move underground between strata as long as there is sufficient pressure to push it upward and sufficient permeable spaces for it to flow through. Oil and petroleum products will tend to move up until they reach a region that traps them in place, with rock formations that are not permeable above them. These regions are known as **traps** and are an essential feature that must be identified in the process of finding oil. Traps come in many varieties, but can be effectively divided into two types: **structural traps** and **stratigraphic traps**. Structural traps are created when a rock formation is warped into a shape that prevents oil from migrating further, such as an anticline or a salt dome (Image 4.17). These traps will be capped with an impenetrable formation above, leading oil to gather in sufficient quantities to be extracted. Because structural traps are relatively easy to identify from a geological perspective, these traps tend to make up most of the productive oil reservoirs worldwide. Unlike structural traps, stratigraphic traps are created by changes in the characteristics of rock formations rather than the shape of rock formations (Image 4.18). Areas of lower permeability or porosity combined with a sealing formation above can create a stratigraphic trap. Because of their greater complexity and inability to be observed by the surrounding rock structures, stratigraphic traps are harder to identify and have only recently become a major source of crude oil and natural gas. New unconventional methods of petroleum exploration and production target oil source rocks where there is essentially no trap; rather, these strata contain oil within tiny pores where it is trapped in place and can only be extracted through horizontal drilling and massive hydraulic fracturing.

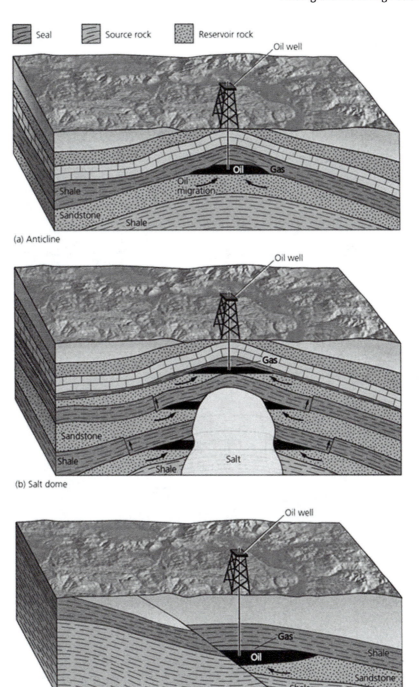

Image 4.17 Oil trapped by geologic structures in (a) anticlines, (b) salt domes, and (c) faults. Natural gas sits on top of the oil reservoir when present.

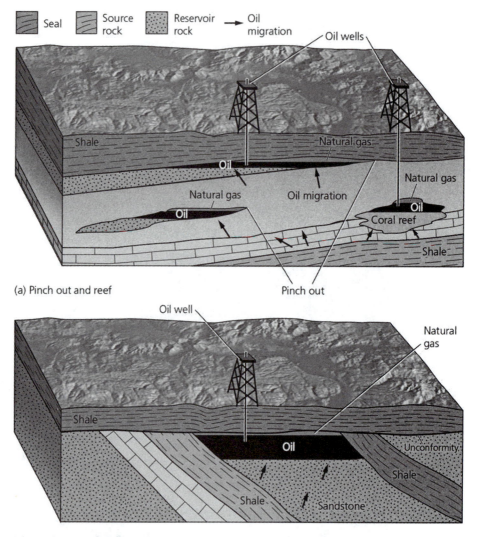

Image 4.18 Oil trapped by stratigraphic features in (a) pinch-out and reef features and (b) angular unconformities. Natural gas will sit on top of the oil when present.

Advances in technology over time have allowed many sources of oil to be exploited that were far beyond the reach of early oil exploration. Drilling advances have allowed much deeper wells than were initially possible, with many exceeding 5,000 meters (16,400 feet) in depth. Another major technological advance has been in the realm of **offshore drilling**, in which deep wells can be drilled by ships or floating platforms (Image 4.19). Some of the most productive wells are produced in offshore locations, but these wells are expensive to drill and involve higher risks. In recent years, offshore

Image 4.19 As technology has developed, offshore drilling has moved into deeper waters.
Eyeidea / Shutterstock

drilling has advanced significantly, allowing wells to be drilled in water over 2.5 kilometers (1.5 miles) deep, with potentially huge production. The environmental risks of offshore drilling are also much greater because it can be difficult to contain a leak or blowout when it is thousands of feet below the water's surface, as happened infamously in the case of the Deepwater Horizon spill in 2010, discussed later in this chapter. Altogether, the methods discussed here comprise the majority of historical oil production. New advances in technology have pushed oil production past these methods, and advances may increase the scope of production yet further.

Unconventional Sources of Petroleum

While most of the world's oil resources have been produced by the methods described in the previous paragraphs, recent technological advances combined with much higher oil prices have opened the field of **unconventional petroleum** exploration. Major new sources of oil, especially in the United States and Canada, are now increasingly obtained through these unconventional sources and technologies, which have greatly changed the nature of the oil industry. In general, unconventional oil comes from rocks that would have previously been uneconomic to produce because of their low permeability. Unconventional sources are those produced with new technologies such as horizontal drilling combined with hydraulic fracturing (fracking) that have revolutionized both oil and natural gas production in the past ten years. Tar sands and heavy oil reservoirs are also considered unconventional although they have been extracted for several decades. At a basic level, oil prices have traditionally been the limiting factor that determined which sources of oil could be economically produced. The rapid rise in oil prices at the turn of the twenty-first century greatly increased the profitability of oil production, making sources that require far more intensive technology or investment to be profitable.

While often associated with natural gas, hydraulic fracturing has been perhaps the most major component of increased oil production from unconventional sources, especially in the United States. Fracturing will be discussed in detail in the next chapter, but it has been instrumental in the rapid increases of oil production in recent years, along with natural gas. Source rocks with low permeability would normally not allow oil to flow to into the oil well and then to the surface, but fracturing and other unconventional treatment methods artificially increase permeability, which then allows these rocks to become an economic producing formation. Some of the largest new oil fields in the world have come online almost exclusively as a result of fracturing technology, which has revitalized oil production from countries such as the United States, which had been in a production decline since the 1970s. Fracturing technologies have been in use since the 1950s, but when combined with horizontal drilling, these methods have revolutionized oil production. By drilling down vertically to a target formation and then drilling horizontally through a long section of that formation, often several kilometers, a much greater amount of producing rock will be exposed to the well bore, greatly increasing production. These horizontal wells are then hydraulically fractured by the pumping of millions of liters of water and millions of kilograms of sand, thus creating a network of permeable fractures that allow these extremely impermeable rocks to produce commercial volumes and rates. Production of *tight oil* with these methods has been a key to the unconventional oil boom in the United States in the past decade.

Other unconventional sources of oil include rocks that require intensive production methods, which often makes them expensive and controversial. Tar sands have recently become a major source of oil for Canada, as discussed later in this chapter, and could become one for Venezuela as well, where the Orinoco tar sands may be the largest remaining oil deposit in the world. Rather than being drilled conventionally, tar sands can be mined for the thick bitumen they contain, which can then be heated to produce crude oil. Similarly, **oil shale** is a type of rock that contains kerogen, the precursor to oil in low-permeability rocks that must be mined and then heated to produce oil (Image 4.20). Much like tar sands, oil shale exists in large quantities around the world, but is expensive to produce and environmentally damaging because of the need for strip mining and enormous amounts of energy to extract oil from the shale rock. Both oil shale and tar sands can also be produced from underground, more like conventional oil, by pumping large amounts of steam into the formation to break down the bitumen, and this process has been used increasingly in the Canadian tar sands (Image 4.21). This process is known as **in situ production**. The only country with large-scale production of oil shale is Estonia, but higher oil prices worldwide could cause other countries with large reserves to increase production, such as the United States and Israel.

Taken together, unconventional oil production methods have revitalized the oil industry worldwide, but remain controversial. The economic benefits of continued petroleum production, along with the continuing demand for oil from developed and

Image 4.20 Certain shales contain vast amounts of oil. These can be fracked or mined and oil can be processed out of the oil shale rocks at the surface.

U.S. Department of Energy/Science Source

Image 4.21 Tar sands, containing large amounts of bitumen, are mined in Alberta, Canada, to produce oil.

Chris Kolaczan / Shutterstock

emerging economies, likely means that production of these unconventional resources will continue for the near future. Major breakthroughs in renewable energy and energy storage technologies could potentially render petroleum economics unattractive, but at present, such major changes remain out of reach.

World Oil Production and Consumption

World production of oil initially began at a large scale in the United States and has been dominated by a group of countries and regions since. Russia (along with several former Soviet republics) has been reliant on oil as its primary export since the 1960s, while production in the Middle East kicked off in 1938 with the first major discovery of oil in Saudi Arabia. The Organization of the Petroleum Exporting Countries, otherwise known as OPEC, was formed in 1960 in an effort to allow major oil-producing countries to cooperate to control prices and demand to a degree, and OPEC member states have become influential through their joint ability to control a resource that is essential to the smooth function of a resource that is so essential to modern life (see Chapter 16 for more information on OPEC).

The Middle East has been the symbol of oil production for decades because several countries in the region are effectively *petro states*, or countries that are almost exclusively reliant on oil and gas for their national economies. Persia (now Iran) was the first such place in which western companies began to produce large quantities of oil, shortly before World War I. However, Saudi Arabia has become the most visible embodiment of a petro state; the country is the largest oil producer in the world (with around 13 percent of total production), it has the largest reserves of conventional oil, and 90 percent of the country's exports are oil or associated products (Image 4.22). Saudi Arabia's Ghawar Field, the world's largest single conventional oil field, produces over 5 million barrels of oil per day, with reserves of around 71 billion barrels. Ghawar Field alone accounts for about 60 percent of Saudi Arabia's entire production since it came online in 1951.

Image 4.22 In 1938, Saudi Arabia first struck oil. It is now one of the world's most important oil producers.

Nick Ludington / Associated Press

Several other Middle Eastern countries are also major oil-producing states, with much of the historical conflict in the region stemming from oil disputes. Iran, Iraq, Kuwait, the United Arab Emirates, Libya, and Algeria (and in the past Bahrain) are all major oil producers, as well as OPEC members (Image 4.23). Conflicts including both Gulf Wars and the Iran–Iraq War have either been driven by oil geopolitics or at least largely financed by oil sales. Other conflicts, such as the rebel insurgency in the Niger delta of Nigeria, have also centered on oil revenues, leading many to consider oil one of the most important geopolitical resources of the modern era.

In more recent years, oil discoveries have brought newfound power and prestige to many other countries around the world, with countries such as Kazakhstan and Venezuela exercising far more political influence than might otherwise be expected as a result of their increasing oil revenues. Despite the greater revenue and influence that oil reserves can bring a country, corruption, conflict over control, and international pressure lead to many petro states like Venezuela or Indonesia having periods of great instability, while others have managed to channel oil money into development and social stability.

As of 2016, world oil production was dominated by three countries, Saudi Arabia, Russia, and the United States, which together comprise almost 40 percent of the global supply (Image 4.24, Table 4.1). Despite their large production volumes, oil plays a different role in each country. Saudi Arabia exports the majority of its production and uses its influence as the largest OPEC exporter to maintain a great degree of regional and global influence. Russia's production comes largely from a handful of state-owned companies operating from declining fields in western Siberia, and it uses oil (along with natural gas) as a major geopolitical tool to influence European countries. Production in the United States from a wide range of private companies largely satisfies the immense local demand, although massive increases in production in the past few years seem set to turn the country back into a net exporter when prices stabilize.

Image 4.23 Oil wealth has allowed cities in oil-producing countries, such as Dubai in the United Arab Emirates, to boom into vast modern cities.
Luciano Mortula / Shutterstock

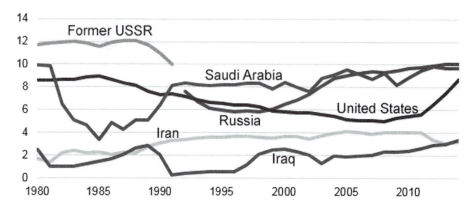

Top five crude oil producing countries, 1980–2016
Million barrels per day

Note: Includes crude oil plus lease condensate. Ranking based on production in 2016.

eia

Image 4.24 Russia, Saudi Arabia, and the United States dominate world oil production.
Courtesy of the U.S. Energy Information Administration

Table 4.1 Oil Production by Country, 2016

2016 Oil production	Thousand barrels/day	Change from 2015 (%)	Share of global total (%)
United States	12,354	−3.2	13.4
Saudi Arabia	12,349	+3	13.4
Russia	11,227	+2.2	12.2
Iran	4,600	+18	5
Iraq	4,465	+10.8	4.8
Canada	4,460	+1.6	4.8
United Arab Emirates	4,073	+3.7	4.4
China	3,999	−7.2	4.3
Kuwait	3,151	+2.7	3.4
Brazil	2,605	+3.2	2.8

Source: Chart from BP Statistical Review of World Energy 2017, https://www.bp.com/content/dam/bp/en/corporate/pdf/energy-economics/statistical-review-2017/bp-statistical-review-of-world-energy-2017-full-report.pdf.

The most recent complete global statistics from 2017 show that the United States has exceeded both Saudi Arabia and Russia to become the world's largest oil-producing country. With the downturn in oil prices, U.S. production may remain stagnant for a time or even decline slightly, so it is possible that these top three spots may shift again, but it is difficult to forecast such changes in advance. For now, the United States, Saudi Arabia,

Table 4.2 Oil Consumption by Country, 2016

2016 Oil consumption	Thousand barrels/day	Change from 2015 (%)	Share of global total (%)
United States	19,631	+0.5	20.3
China	12,381	+3.3	12.8
India	4,489	+7.8	4.6
Japan	4,037	−2.5	4.2
Saudi Arabia	3,906	+1	4
Russia	3,203	+2.1	3.3
Brazil	3,018	−4.8	3.1
South Korea	2,763	+7.2	2.9
Germany	2,394	+2.3	2.5
Canada	2,343	+1.9	2.4

Source: Chart from BP Statistical Review of World Energy 2017, https://www.bp.com/content/dam/bp/en/corporate/pdf/energy-economics/statistical-review-2017/bp-statistical-review-of-world-energy-2017-full-report.pdf.

and Russia are set to remain by far the largest oil producers, regardless of fluctuations relative to each other. Table 4.1 further shows that many traditional oil powers in the Middle East, such as Iran and Iraq, have regained their leading status as global exporters after the lifting of sanctions and the decline of ISIS, respectively. Other states outside the Middle East such as Canada and China continue to play important roles in global production as well. Volatility in the global oil market will ensure that production changes between regions continue into the future as new technology opens previously inaccessible resources, while older fields decline until they can no longer be produced economically.

Global oil use paints a different picture than that of production, with the top industrial powers of the world consuming vastly more oil than other countries (Table 4.2). China and the United States together account for almost one-third of all oil consumption worldwide. While growth in the United States has slowed to a degree, Chinese industrialization has continued to accelerate demand for oil, particularly with the massive increases in car ownership. Other developed countries show a similar slowdown because population growth has slowed down (or stopped entirely) and efficiency standards have greatly reduced per capita energy use. Developing countries like India have experienced large increases in demand, while others such as Brazil have grown at an inconsistent pace. Altogether, the industrialized world has seen a general flattening of demand, primarily because of efficiency improvements, while the increasing global demand has been driven by less developed economies that are experiencing booms in energy demand and car usage.

Oil Production and Consumption in the United States

As the first country to begin extraction of oil at industrial levels and the first to pioneer oil drilling technology, the United States has been a major player in the world oil market since its beginning. Additionally, as the largest economy in the world during the petroleum era, the United States has been the world's primary consumer of oil. While it has been a major consumer and an important producer, American oil production was not sufficient to supply its own needs after the industrialization of the country gained speed in the mid-twentieth century, leading to the need for imports. Oil production in the United States peaked in 1970, and imports of foreign oil surpassed domestic production in the mid-1980s, making the country reliant on external producers. With the advent of advanced drilling technologies such as horizontal drilling and advanced fracturing techniques, U.S. oil production has reversed its decline for the first time since 1970, making domestically produced petroleum once again a major player in the domestic energy market. In 2013, the United States imported less than half the oil it consumed for the first time since 1987, marking a major swing from the trend of increasing demand and decreasing local supply.

Oil production in the United States began in Pennsylvania, but quickly moved to the west, particularly Texas and California. Since then, these states have remained major players in the United States and have been joined by a few other geographic areas opened by new technologies. While Texas is still by far the largest producing state, North Dakota has experienced a massive boom in the Williston basin, primarily a result of advances in unconventional technologies. These advances have opened other massive projects to exploitation, such as the Eagle Ford shale play in South Texas, which yields over 1.5 million barrels of oil per day, well over 10 percent of the entire U.S. production (Table 4.3). The massive increases in oil from unconventional fields in the United States, combined with a global economic slowing in energy demand that caused the major decline in oil prices in late 2014, will likely continue to drive a large degree of price volatility in the future. While these price drops put downward pressure on the domestic oil industry, the production increases have also greatly reduced national dependence on foreign sources of oil, long a political goal of American leaders.

In addition to this boom in Texas, North Dakota, and Alaska, offshore drilling has become a major part of U.S. oil production, particularly in the Gulf of Mexico (Table 4.4). Taken as a region, the offshore Gulf of Mexico produces 1.6 million barrels of oil per day, second only to Texas. Offshore oil and gas extraction can yield impressive results because the immense pressure of offshore wells allows them to produce much faster than onshore wells, but operating at depths over 2 kilometers (1.2 miles) raises the cost and

Table 4.3 Largest Oil Fields in the United States by Production as of 2013

Largest U.S. oil fields	Location
Eagleville	Texas
Spraberry Trend	Texas
Prudhoe Bay	Alaska
Wattenberg	Colorado
Briscoe Ranch	Texas

Source: Energy Information Administration, Top 100 U.S. Oil and Gas Fields, https://www.eia.gov/naturalgas/crudeoilreserves/top100/pdf/top100.pdf.

Table 4.4 Oil Production by State/Region in 2016, Thousand Barrels per Day Average

State/region	Thousand barrels/day	Change, 2015–2016 (%)
Texas	3,213	−6.8
Offshore Gulf of Mexico	1,598	+5.5
North Dakota	1,033	−12.2
California	508	−7.8
Alaska	490	+1.4
Oklahoma	420	−6
New Mexico	399	−1.2
Colorado	317	−5.7
Wyoming	198	−16.5
Louisiana	154	−10.5

Source: U.S. Energy Information Administration, "Crude Oil Production," https://www.eia.gov/dnav/pet/PET_CRD_CRPDN_ADC_MBBLPD_A.htm.

risk of failure commensurately. The Deepwater Horizon spill is a particularly stark example of the risks of offshore drilling, but the large payoff for success is likely to ensure that offshore oil remains a major part of U.S. petroleum resources. Because of the massive breakthroughs in onshore unconventional production, however, offshore is often less attractive for new investment and has declined during periods of sustained low oil prices. Because onshore has such low risk and investment needs in comparison, offshore wells can only be drilled by the largest companies with the greatest financial and technical resources.

Until the recent downturn in world oil prices, it appeared that the United States might continue to increase its oil production and eventually become a net oil exporter. The high costs associated with many unconventional fields have dampened the explosive growth of U.S. production and it remains to be seen where domestic oil output will be in the next few years.

VIGNETTE
Canada's Tar Sands

The move toward unconventional sources of oil in North America has boosted production and revitalized the oil industry, yet some of these methods have generated substantial controversy for their environmental impacts. Of all new sources, none is as controversial as the tar sands of Canada, which are widely perceived to be one of the dirtiest methods of energy extraction currently in use. Despite their large environmental impact, the sands represent a major source of oil for Canada and its energy companies,

continued

continued

presenting an economic and geopolitical benefit that is proving difficult to resist at a national level. Controversy over the sands has spilled over into the United States, with the Keystone XL pipeline, which aimed to pipe crude oil from Canada to the U.S. Gulf Coast (Image 4.25), serving as a rallying point for environmental groups to call for a move away from tar sand production. In reality, opposition to the Keystone pipeline was more of a symbolic protest against tar sands oil than a practical solution; oil transported by the pipeline would have represented less than 1 percent of U.S. carbon emissions, and the same oil would eventually be shipped by rail or another pipeline to a different buyer to be consumed in the end. However, the pipeline debate demonstrates how international politics, energy security, and environmentalism have all become intertwined in the modern energy debate, all of which helped lead the U.S. government to deny the pipeline permit in 2015, with a presidential order reversing the decision and granting approval signed in 2017.

While the final product of tar sands is similar to that of other crude oil products after refining, it takes far more energy and processing to get to that point. Rather than being drilled for, most tar sands oil is currently strip mined from shallow deposits that are rich in bitumen, a much thicker form of crude oil that resembles tar in

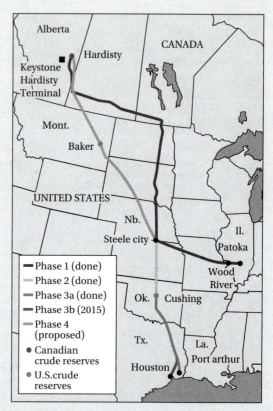

Image 4.25 The proposed Keystone XL pipeline would allow for the transportation of oil from Canada's tar sands to refineries on the U.S. Gulf Coast. User:Meclee / Wikimedia Commons / CC-BY-SA-3.0

consistency. Each barrel of tar sands oil requires the removal of 2 to 4 tons of earth from above it, and it must then be processed on-site before further transport. The bitumen, once separated from its source rock, must be "cracked" with large amounts of heat and water to turn it into a transportable form of crude oil. This process requires large heat inputs in the form of burning natural gas and produces sizeable impacts. Large amounts of greenhouse gases are released during the heating process, and huge amounts of waste rock, bitumen, and polluted water are also created. These waste products are generally stored in massive collection ponds that are open to the air, which have resulted in the deaths of thousands of migratory aquatic birds that land in them. Further, the large scale of surface mining leaves the local landscape scarred, open to erosion, and likely to seep waste products into local water sources (Image 4.26). There are also worries that these pond levies could breech and contaminate nearby rivers, especially the Athabasca River (Image 4.27).

Image 4.26 Development in the tar sands has led to vast areas of deforestation, along with large tailing ponds filled with toxic waste products from tar sands production.
Jeff McIntosh / Associated Press

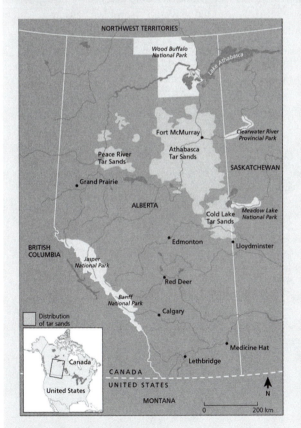

Image 4.27 Locations of the three largest tar sand deposits in Alberta, with the Athabasca deposit being the largest.

The sum of environmentally damaging aspects of the tar sands has made them politically contentious, but the economic benefits for Canada have so far outweighed any environmental pressure in the domestic debate. Regulations or taxes on carbon emissions could adjust the economic attractiveness of the tar sands in favor of promoting other, more sustainable energy resources, but resistance to these types of rules from energy-exporting countries and industrial groups has been strong. Some minor carbon capture and sequestration operations have been implemented on a small scale to test the feasibility of reducing carbon emissions from the tar sands, although these are purely experimental at present. As long as the economic benefit for Canada and its companies continues to be lucrative and a major provider of energy security, it is likely that tar sands will continue to be exploited despite pressure from environmental groups and the ongoing degradation of Canada's natural environment.

Beyond Peak Oil

Oil takes millions of years to form and is relatively rare in its occurrence and accessibility. Because of the timeline on which oil forms, it is effectively not being created at a rate that makes it renewable for humanity. Consequently, for practical purposes, the amount of oil that exists is finite. As more oil is produced, the total global amount is depleted, causing prices to rise and limiting further oil resources to places from which it is harder and more expensive to extract. These basic observations led to the development of **peak oil theory**, developed by M. King Hubbert in 1956. According to Hubbert's observations, continuous production of oil would lead to an eventual maximum rate of total production, or peak, after which production would decline permanently. Oil production after the peak would be in more difficult-to-reach places or be more expensive to produce, making further production more and more difficult, which would in turn cause prices to increase. Eventually, most accessible oil on the earth will have been extracted, effectively making the supply drop to zero. In practice, the most expensive and inaccessible oil would likely never be extracted. Based on the trends Hubbert observed, this production would resemble a bell curve, which is why this circumstance has come to be known as a **Hubbert curve**, or Hubbert's peak.

Advocates of the peak oil theory have pointed out that, until recently at least, world production strongly followed the Hubbert curve, leading many to use peak oil as a predictor of future oil production (Image 4.28). If peak oil production holds true, the

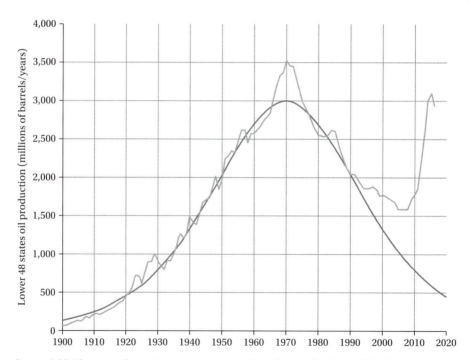

Image 4.28 The recent boom in unconventional oil production has led to U.S. oil production moving away from the predicted Hubbert curve.

increasing scarcity of oil combined with the increasing global demand for it could lead to catastrophic consequences. As production drops while demand increases, prices will rise rapidly, as seen in the early 2000s. Without the development of alternative fuels to replace oil, the rapidly declining supply could lead to severe economic impacts and global shortages because so many aspects of modern society depend on a regular and predictable supply of oil. Predictions of when this peak will occur have frequently been made and revised with newer data, but widespread debate on the relevance of peak oil theory remains. While oil is indeed a finite resource, peak oil theory did not forecast the development of unconventional oil sources, which have since caused increased production levels in certain countries. If applied worldwide, these technologies could potentially reverse the global declines in oil production, at least for a discrete period of time. The underlying limited supply of oil will still mean that the resource will eventually run dry, but forecasting this date becomes more difficult as newer sources begin to be exploited.

A key aspect to the peak oil debate centers on oil's place in society in the coming decades. While it is difficult to forecast future changes in the overall energy basket, a key component to peak oil is whether viable replacements can be found for it. Because oil is used for so many different applications, this further complicates any assumptions that can be made about future demand. If battery technologies are sufficiently advanced in the coming years, oil's dominance as a transportation fuel could come to an end, but this would not impact its essential role as a feedstock for plastics and other industrial products. Many detractors of peak oil theory point to these uncertainties as weaknesses of the theory's predictive abilities. What is clear, however, is that the supply of oil is finite and, on a certain timeline, is likely to either run out almost completely or be abandoned altogether. Because of the many interdependent factors that will determine how we use (or do not use) oil in the future, it will be impossible to say when, or whether, we hit peak oil for the near future.

Impacts of Oil Production and Consumption

Oil exploration and production are fraught with environmental impacts, like all fossil fuels. Oil is a major contributor to many forms of pollution and environmental damage, most important as the base of gasoline and diesel air pollution from transportation, but also through damage during drilling, oil spills during extraction, transport and storage, and its final contribution of chemicals that create acid rain, smog, and greenhouse gases after combustion. These environmental impacts comprise the primary argument for reducing global dependence on oil as an energy resource, although the overwhelming dominance of oil in the current global energy system means that it will continue to be a major source of environmental concern for some time.

The environmental impacts of oil begin with the drilling process, which can range from minimally invasive to highly disruptive. Conventional onshore oil wells require little disturbance of the surrounding landscape, aside from the construction of a basic drilling pad, which houses the drilling rig and equipment. In more fragile environments like the Arctic, even these disturbances can be significant. One of the crucial elements of drilling is the disposal of wastewater from the well itself; most oil reservoirs contain large amounts of water that come to the surface as a byproduct along with oil. This water contains a large quantity of contaminants from the reservoir, making it unsuitable for other applications, and it must be disposed of. A common solution is to inject it back into the original reservoir or other deep rock formations that would be sealed and incapable of allowing it to return to the surface. When subsurface disposal is not an option, however, it must be transported to special water treatment plants that remove the contaminants.

Along with water disposal, there is a possibility of spills and leaks from oil wells. With onshore drilling, these spills and leaks are generally minor and easy to plug before major damage is done. Offshore leaks can be much more difficult to seal, leading to the possibility of such catastrophes as the Deepwater Horizon spill of 2010. Hydraulic fracturing has also garnered attention in recent years as a source of potential environmental damage, but controversy surrounds the scientific validity of many of the claims. Fracturing is discussed in greater detail in the next chapter. Last, certain methods of oil exploitation have much larger environmental footprints, especially those requiring open mining operations, such as the tar sands of Alberta. While these methods do not currently represent a major part of worldwide oil production, they still account for a significant amount of the environmental impact of oil production.

While drilling may be the most visible aspect of oil production, its greatest environmental impacts come after it has been extracted from the ground. Oil spills are a visceral representation of the type of damage oil can do if released directly into the environment. Major oil spills result from oil tankers floundering or sinking, which can release thousands or even millions of barrels into oceans, rivers, or lakes. In marine environments, oil is extremely toxic to fish and other marine life and can do serious damage to fishing industries, as well as creating unsightly shoreline pollution that can harm humans and coastal wildlife. Other spills can result from land-based transport of oil, such as ruptured pipelines or crashes of trucks or trains with oil transport tanks (Images 4.29–4.32). In general, pipelines are the safest way to transport oil

Image 4.29 Pipelines are the most efficient and safest means of transporting oil long distances.
BILD LLC / Shutterstock

Image 4.30 Oil tankers carry crude oil or refined products overseas.
Teun van den Dries / Shutterstock

Image 4.31 Oil trains offer another means to move oil overland besides pipelines, but several major derailments have led to major oil spills and explosions.
Satephoto / Shutterstock

because they utilize technology that initializes automatic shutoffs in case of emergency, greatly reducing both the amount of spilled oil and the geographic area that it can contaminate. Rail- and road-based transport of oil is less reliable and is usually only used when pipelines have not yet been built or have been politically opposed for other reasons.

Image 4.32 The *Exxon Valdez* oil spill in Alaska galvanized attention to the dangers of oil transportation. Large amounts of marine life were killed by the spill and pristine areas along Alaska's coastline were damaged.
Rob Stapleton / Associated Press

While drilling, oil spills, pipelines, and transport accidents make for the most visible symbols of oil's impact on the environment, the combustion of petroleum products at their end uses represents by far the largest environmental impact, ranging from local air pollution damaging health, to acid rain, to being a primary driver of human-induced climate change. Petroleum combustion is a major contributor to local and regional air pollution because the incomplete combustion of petroleum creates a wide variety of chemical compounds. Carbon monoxide is a major air pollutant, and particulate pollution is also a significant health hazard in large concentrations because these compounds cause numerous lung and heart problems in highly polluted zones. The combustion process of petroleum, like that of other fossil fuels, also creates significant amounts of nitrous oxides and sulfur dioxide, which are the chief contributors to acid rain. Acid rainfall damages local forests, marine ecosystems, and coral reefs and causes physical damage to urban structures. Further, oil burning produces ozone, which creates smog when interacting with sunlight (Image 4.33).

Climate change has been driven by many human processes over the past century, and the combustion of oil and petroleum products has been one of the largest contributors, primarily through the release of large amounts of CO_2. Coal has overtaken oil as the largest global producer of atmospheric CO_2 in the past decade, but oil had been the largest for several decades prior, and it continues to be close behind coal in total emissions by mass. Transport fuel is the largest component of global CO_2 emissions, comprising 22 percent of all emissions globally. The majority of transport fuel is supplied by oil products such as gasoline, diesel, and jet fuel, and the continued growth in demand for transport fuels is forecasted to increase the amount of greenhouse gas emissions sourced from oil. In the United States, emissions from oil have fluctuated, but have not increased markedly since 1971, with 2011 emissions being lower than those in 1980 (Image 4.34). Stricter emissions regulations and the development of better combustion technologies such as catalytic convertors have helped to mitigate the growth of oil-based emissions despite continued population growth and a vast increase in the amount of vehicles on American roads. Increased fuel efficiency has also aided in reducing the growth of emissions and is estimated to be the biggest factor in the flattening of American CO_2 emissions from oil.

Image 4.33 The smog in Los Angeles, California, is primarily caused by vehicle emissions burning gasoline and diesel along the city's famously congested roads.

Andrius K / Shutterstock

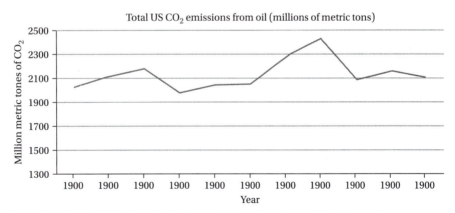

Image 4.34 Total U.S. CO_2 emissions from oil. Data from International Energy Agency, *CO_2 Emissions from Fuel Combustion*, 2013 edition, http://www.iea.org/w/bookshop/add.aspx?id=633.

Deepwater Horizon Oil Spill

Oil spills are some of the most visible manmade environmental disasters and have come to represent the ecological impact of the oil industry as a whole. Some spills can be so large and damaging that they result in major public campaigns and government shifts in policy, often aimed at accelerating the reduction in dependence on oil. The *Exxon Valdez* spill in 1989 remained the largest spill in U.S. history (up to 750,000 barrels spilled after a tanker ran aground in Alaska) until the now-infamous Deepwater Horizon spill of 2010. Deepwater Horizon was an offshore drilling rig operating in the deep-water sections of the Gulf of Mexico south of Louisiana, a particularly expensive and difficult region to produce from. During the drilling process, the failure to adequately monitor the downhole pressure resulted in a massive blowout, in which the pressurized fluids in the well bore are forced back out at the surface in an explosion. Usually this type of accident would be prevented by a mechanical blowout preventer, but the device on Deepwater Horizon failed critically, allowing the blowout to proceed.

Once the flammable fluids from downhole reached the surface in massive concentrations, they ignited, destroying the drilling platform and killing eleven workers (Image 4.35). At the seabed surface, the failed equipment allowed oil to escape into the seawater at an uncontrolled rate, well over 50,000 barrels per day. Efforts to seal the well and prevent further release repeatedly failed; the well breached on April 20 and was not stopped until mid-June; the well was not completely sealed until early September. In total, the well released almost 5 million barrels into the Gulf of Mexico, making it one of the largest oil spills in history. Cleanup efforts have taken years and the public health and environmental impacts of the spill will continue for many years to come (Image 4.36). BP, the well's primary operator, along with Transocean, Halliburton, and other involved contractors, has been made to pay tens of billions of dollars in cleanup costs, with further legal action pending. While oil production in the United States, both offshore and onshore, continues to rise, events like the Deepwater Horizon spill continue to increase public pressure to move away from fossil fuel use.

Image 4.35 Deepwater Horizon, a large offshore oil rig, caught fire and exploded, killing workers on the rig and triggering a massive blowout and oil slick in the Gulf of Mexico that devastated the coastal economies of Louisiana, Mississippi, Alabama, and Florida.
Gerald Herbert / Associated Press

Image 4.36 The oil spill from the Deepwater Horizon blowout killed and injured large numbers of marine life, such as this pelican.

Charlie Riedel / Associated Press

REVIEW QUESTIONS

1. Are any of the ancient uses of crude oil or bitumen still used today?
2. What are a few oil-derived products that you regularly use other than gasoline?
3. Why is oil only formed and found in sedimentary rocks?
4. Why must crude oil be refined, rather than just consumed as it comes out of the ground?
5. Other than gasoline, what are the primary products refined from crude oil? Which are lighter than gasoline and which are heavier by volume?
6. Are oil traps indicative of where oil forms in rock or just where it travels after formation?
7. What types of oil production are unconventional?
8. Although the United States imports oil, it also is a major producer. Is it likely to remain an importer if oil prices rise?
9. Will we ever totally run out of oil?
10. Why were environmentalists so opposed to the Keystone XL pipeline? Will the denial of a permit stop tar sands production in Canada?
11. Beyond oil spills, what are the other major environmental impacts of oil production, transport, storage, and consumption?

Image 5.1 Pi-Lens / Shutterstock

CHAPTER 5

Natural Gas

Natural gas, unlike coal and oil, is a relatively modern energy source. While the substance itself was known to many cultures around the world, the Chinese were the first to harvest the resource directly. By tapping gas seeps with shallow wells, natural gas was extracted and piped through bamboo tubes to be burned in the production of sea salt by the Chinese over 2,500 years ago. Apart from this and few other minor uses over the following centuries, no other cultures attempted to harvest natural gas for any significant use until the Industrial Revolution. Even after oil drilling was developed in the West, most gas associated with oil was flared or vented to the atmosphere because it was essentially a waste product with no market. Only after pipeline and storage systems for gas were devised and widely implemented did gas become a product with commercial value.

Natural gas, although it is distinct from oil, is also a nearly ubiquitous byproduct of oil production because it is dissolved in oil at subsurface pressures and released at surface pressures, similar to carbon dioxide (CO_2) in soft drinks. This type of gas is referred to as *associated gas* and is distinct from nonassociated or *dry gas*, which is produced without liquid hydrocarbons. Both gas types are drilled and developed using the same techniques as for oil production, but natural gas must be handled differently at the surface and for its distribution. Oil can be stored by any method appropriate for liquids, while natural gas requires a range of more sophisticated equipment to separate, transport, and store, with gas pipelines usually being the cheapest transit solution. Consequently, the use of natural gas did not peak along with that of oil, only emerging as a major energy source well into the twentieth century.

In general, natural gas is less versatile than oil because it cannot be refined into as many products or transported as easily, but it also has several advantages that have made it one of the dominant fossil fuels in the modern era. A primary benefit of natural gas is that it is plentiful since the range of geologic conditions favorable for natural gas is greater than that for oil. Natural gas also has the benefit of being the cleanest fossil fuel to burn because it is chemically richer in hydrogen and contains fewer impurities and extraneous compounds that are released during combustion. For electricity generation, this makes gas a much cleaner alternative to coal- or oil-burning power plants, producing 30 percent lower greenhouse gas emissions than oil and 45 percent lower emissions than coal to generate the same amount of electricity, while also dramatically reducing the emission of other air pollutants such as sulfur and particulates. Clean-burning natural gas is ideally suited to home heating and cooking, which remain some of its most important uses (Image 5.2). It is also used widely as a feedstock for fertilizers.

Natural gas usage in the United States in the twenty-first century has been soaring as a result of rapidly expanding reserves as drilling and fracturing technologies have dramatically increased the gas supply. A direct result has been a significant drop in the price of gas compared to oil (Image 5.3). In the recent past, natural gas prices were closely tied to oil on an energy-equivalent basis, that is, 1 million British thermal units of natural gas would cost approximately the same as 1 million British thermal units of oil. The decoupling of gas from oil prices has had a pronounced effect in North America, but prices of natural gas vary significantly by world region, whereas crude oil prices are far more even globally. While the United States and Canada have well over one hundred years of gas reserves at current usage rates, many other regions have almost no gas reserves at all; consequently, their prices are many times higher because of the high transit costs for natural gas compared to oil.

Image 5.2 We use natural gas in many scales, from electrical production and industry to home uses for heating and cooking.
Roman Sigaev / Shutterstock

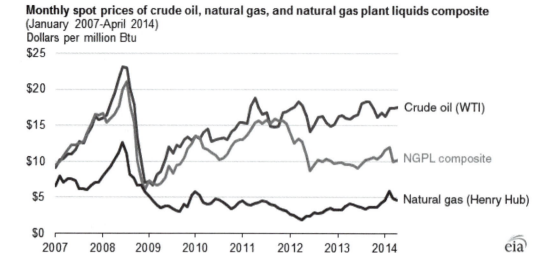

Monthly spot prices of crude oil, natural gas, and natural gas plant liquids composite
(January 2007-April 2014)
Dollars per million Btu

— Crude oil (WTI)

NGPL composite

Natural gas (Henry Hub)

eia

Image 5.3 In recent years, the prices for natural gas and natural gas liquids have decoupled from oil prices in the United States.

Courtesy of the U.S. Energy Information Administration

Since 2006, as unconventional gas technologies increased the U.S. supply, natural gas has become significantly less expensive than oil on an energy-equivalent basis and has been displacing coal in power generation, expanding its role beyond traditional home heating and industrial use. In addition, natural gas also has the potential to be used as a transportation fuel when compressed. Its lower price and cleaner combustion make this use an attractive possibility for the future, but even when compressed, it is less energy dense than gasoline, necessitating larger fuel tanks in cars and trucks. Many have suggested that natural gas can serve as a transitional fuel (replacing coal and oil temporarily) while renewable technologies are scaled up.

In this chapter, we will learn about the origins and various reservoirs for natural gas, its uses as an energy resource and chemical feedstock, the means by which it is produced—including through hydraulic fracturing—and its role in the global and U.S. economies. We will also explore nontraditional sources of methane and investigate the various environmental concerns associated with natural gas.

Origins and Geology of Natural Gas

Natural gas is much like oil, not a single substance, but a mix of several naturally occurring compounds (Table 5.1). Natural gas is primarily composed of **methane**, with lesser amounts of **ethane**, **propane**, **butane**, hydrogen, and sometimes **pentane**. The heavier associated gases that occur with methane are generally separated and used as their own fuels after extraction. Once the heavier gases are separated out, the remaining gas, almost exclusively methane, is ready for transport into a consumer network such as a utility.

Natural gas is created by two primary processes: biogenic and thermogenic formation (Image 5.4). **Biogenic methane**, or **biogas**, is mostly formed in shallow rock or

Table 5.1 Components of Natural Gas

Compound	Chemical formula	Uses
Methane	CH_4	Domestic and industrial fuel, feedstock for fertilizer
Ethane	C_2H_6	Feedstock for the production of ethylene
Propane	C_3H_8	Fuel for small engines and heating uses
Butane	C_4H_{10}	Fuel for gasoline refining, chemical feedstock, refrigeration, and lighter fluid
Pentane	C_5H_{12}	Agent for foam production, industrial solvent
Hydrogen	H_2	Processing agent in chemical and petroleum production, coolant, energy storage medium

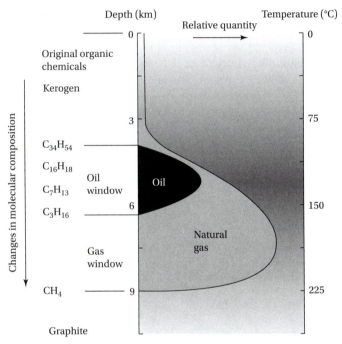

Image 5.4 Natural gas forms in a much larger window of temperature, pressure, and time than crude oil. Natural gas also forms in the oil window, along with crude oil.

Information from University of California, San Diego

coal seam reservoirs by microbes that consume organic material within the rock and produce methane as a waste product. Other large deposits of dense organic material, such as swamps, landfills, or sewage storage areas, can also produce methane through biological activity, and most of this gas is emitted directly into the atmosphere. In some places, biogas is collected for use as a fuel, but the large amounts of CO_2 that are usually associated with it mean that some processing is normally required before it can meet acceptable emissions and pipeline standards. Biogas could be a more significant source of energy worldwide if the infrastructure to capture it were built, and further capture would have the added benefit of treating the gas before use, thus lowering the emissions of CO_2, methane, and other gases that would normally be directly released into the atmosphere (see Chapter 10 on biofuels).

The majority of the natural gas used worldwide is **thermogenic natural gas**, which is older and generally richer in methane than biogenic gas. Thermogenic natural gas is created in much the same way as oil, through high heat over millions of years. Organic compounds are buried deep in the earth through geologic processes, where heat transforms them into kerogen, the precursor of hydrocarbons. Kerogens vary in type depending on the kind of organic material they are derived from. Some kerogens are ideally suited to oil generation, while others tend to convert almost exclusively to natural gas. As kerogens become more deeply buried and heated to higher temperatures, thermal conversion to gas occurs. Even for types of kerogen that generate oil, high enough temperatures will eventually cause a conversion entirely to natural gas. Natural gas produced along with oil is known as **associated gas** and is often produced because the more valuable oil being extracted happens to contain quantities of gas as well. Before the infrastructure to capture associated gas was developed, associated gas would usually be burned off (known as flaring) rather than sold (Image 5.5). In some places where there is still insufficient infrastructure to capture associated gas, such as Nigeria, flaring is still routinely practiced. **Unassociated natural gas**, or gas with no oil in the same reservoir, makes up the bulk of gas production worldwide.

Natural gas can be found in almost all the same types of geologic formations as oil, as well as some unique occurrences, such as **coal-bed methane** (Image 5.6). Coal-bed methane is extracted from coal seams, where natural gas has filled the spaces in the

Image 5.5 At many oil wells, natural gas is burned off, or flared, rather than captured for consumption or reinjection into the well. This wastes a valuable resource, while simultaneously contributing carbon dioxide to the atmosphere.
HeliHead / Shutterstock

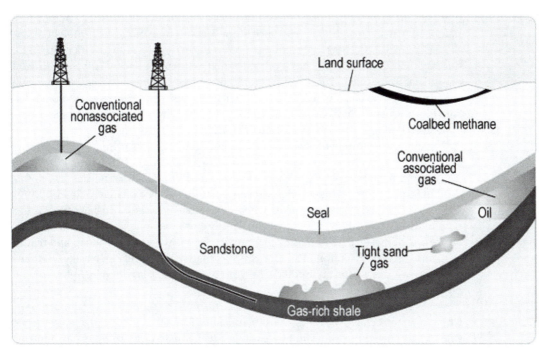

Image 5.6 Natural gas is produced from a variety of formations including conventional deposits, tight and shale gas deposits, and coal-bed methane.
Courtesy of the U.S. Energy Information Administration

highly porous matrix of coal. Unlike traditional natural gas, coal-bed methane usually comes from shallow wells, and it contains low amounts of heavier gaseous hydrocarbons such as butane. Because it is produced more unconventionally, coal-bed methane has only been extensively produced in North America and Australia, although significant reserves exist elsewhere. It can be produced in areas where coal is mined, sometimes to reduce the risk of explosion in underground coal seams, but also from coal that is too deep or of too poor quality to economically mine.

Because natural gas forms under a wider variety of conditions than oil, it is more common than oil worldwide. Additionally, natural gas is often distributed in different geographic regions from oil, so some countries that have comparatively small oil reserves may be richer in gas, making it politically and economically separate from oil. As will be discussed later in this chapter, the abundance of natural gas has the possibility to make it a more important resource, but there are also many challenges to be faced logistically, politically, and environmentally.

Uses of Natural Gas

As one of the major fossil fuels, natural gas has a range of uses in many different industries. Apart from its use as a feedstock for fertilizers, natural gas is not generally refined into other products, limiting it mostly to combustion-driven uses. Because of its pure nature, natural gas can be used directly in a range of applications from domestic to industrial, and the lack of major processing or refining needs is one of its key strengths. Because of the direct combustibility of gas, it has a versatility that neither oil nor coal enjoys; while coal, oil, and natural gas can all be used industrially, in electricity generation, and domestically for heating and cooking, natural gas burns relatively clean and can be used more safely with less processing than coal and oil. This is particularly important for domestic use because wood, coal, and oil stoves are much more hazardous to the health of their users.

The domestic burning of natural gas has long been one of its primary uses and continues to be a driver for natural gas consumption in many parts of the world. With an integrated municipal supply of gas lines to houses and commercial properties, gas is often used in furnaces, stoves, and boilers. Because natural gas is a relatively clean-burning fuel, it is well suited for use in houses, where coal had been a much dirtier domestic fuel before the widespread availability of gas. Winter home heating and water heating are the largest domestic uses of natural gas, and in regions where gas is cheap and readily available, gas is more common as a heating fuel than heating oil or electricity. Methane is an odorless gas, so the odor noted when there is a gas leak is added by utility companies for safety reasons.

Gas can also be used by industry to power turbines or other industrial machinery, where it is both cheaper and cleaner than coal or fuel oil. Gas can be used to power smelting furnaces and other similar heat-intensive industrial processes, where the high heat potential is more efficient than direct electricity. Natural gas is also used in the industrial production of nitrogen fertilizers, where it is an important feedstock that is reformed through the Haber process (Image 5.7) to produce ammonia; this use represents one of the few noncombustion applications of gas (Image 5.8).

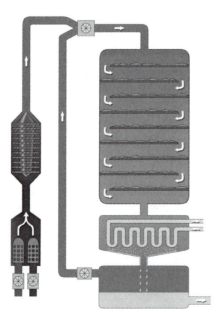

Image 5.7 In this diagram of the Haber process, ammonia is produced from nitrogen and hydrogen gas. Hydrogen (red) and nitrogen (blue) are first cleaned (lower left), mixed, and then compressed (center left). The mixture cycles through the reaction tower (right), and a catalyst (iron, on horizontal trays) and heat form ammonia. A cooling loop (water, pale blue) at the base of the reaction tower condenses the ammonia into a liquid (orange) that is piped off at lower right. Courtesy of SPL / Science Source.

Science Photo Library / Science Source

Along with these industrial uses, **compressed natural gas** (CNG) has been increasingly used as a transport fuel, most commonly in city buses (Image 5.9), but also in cars in regions where there is sufficient infrastructure, such as parts of California. If natural gas prices remain low for many years and oil prices remain high, CNG may continue to grow

Image 5.8 Natural gas is an important feedstock in the production of fertilizer. It therefore plays an important role in ensuring enough food to meet our needs.
Oticki / Shutterstock

Image 5.9 Although far less common for transportation fuel than oil-based products, natural gas is commonly used for buses in cities. It helps reduce local air pollution compared with diesel or gasoline engines.
Ashley Cooper/Science Photo Library

Image 5.10 Combined-cycle natural gas power plants turn turbines both from the flow of gas into the plant and again in the conventional manner using steam once the gas is burned in a boiler.
Peter Bowater/Science Photo Library

as a transport fuel, although the parallel development of electric cars will be an important influence on whether the transport industry adopts one or both alternative technologies.

In recent years, natural gas use has experienced the greatest growth in the generation of electricity, especially in the United States and other developed countries. While natural gas has been used in electricity generation for decades, the recent growth in its supply, low cost, and environmental advantages over coal have driven a major shift away from coal as the dominant electricity source in the United States. Newer natural gas power plants have achieved a high level of efficiency by utilizing combined-cycle technologies. A combined-cycle gas generator burns natural gas in a turbine and then uses the hot exhaust gases from the turbine to run a second steam or other working fluid generator (Image 5.10).

Finding and Producing Natural Gas

Given that natural gas is largely found in the same sedimentary geologic regions as oil, it is often produced in much the same way. Geologists must first locate economical natural gas reservoirs using tools similar to those used in oil exploration, and then the reservoirs can be drilled using essentially the same technologies. Many of the same new technologies that have been used successfully to produce unconventional petroleum resources (see Chapter 4) have also been used to produce natural gas, and most energy companies explore for and produce both products. Many reservoirs contain both resources in the same place, so wells that are targeted to produce oil will often produce natural gas as a byproduct or vice versa. The primary difference between oil and natural gas production comes in the transport and usage phases of the process because liquids are easier and cheaper to transport, especially when shipping overseas. These differences, both in producing regions and in the infrastructure of gas transport, have contributed greatly to the vast differences in the economics, geopolitics, and environmental impacts of the natural gas industry versus that of oil.

As with oil, the production of natural gas has been greatly changed in recent years by the *unconventional revolution*. Horizontal drilling and improved hydraulic fracturing (fracking) technologies have combined to allow gas production from shale and other reservoir types that were once impossible to exploit (Image 5.11).

Conventional production of natural gas works in much the same way as conventional oil drilling, in which a vertical well is drilled through the target formation, where subsurface pressure then allows gas and condensates (liquid hydrocarbons often associated with natural gas) to flow or be pumped to the surface. Before horizontal drilling, vertical wells went nearly straight down, and the well consequently only was exposed to the reservoir for a few meters. With new drilling technologies, wells are drilled to the target formation, and then the drill is turned to penetrate the rock unit horizontally,

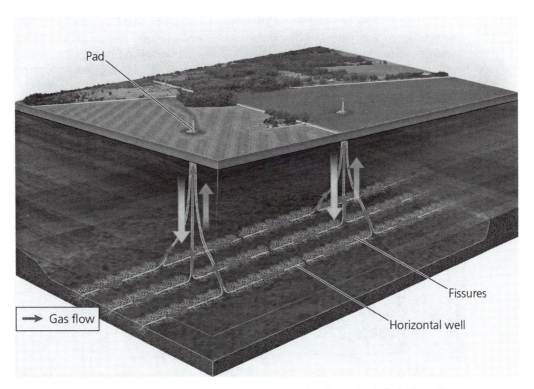

Image 5.11 From a single pad, multiple horizontal wells can be drilled. The wells turn horizontal at the target formation and run along it, allowing much greater contact with the pipe to increase production.

in which it can continue to be drilled for thousands of meters. In this way, far more of the target rock will be directly exposed to the well. In shales and other less porous rocks than sandstone, the horizontal section of the well will then be fractured with fluid (primarily water) and **proppant** (usually sand) at high pressure to allow hydrocarbons to flow. Hydraulic fracturing is discussed in detail in this chapter in "The Hydraulic Fracturing Controversy."

As an economic commodity, several factors complicate the supply and production of natural gas on a global scale. Because it is a gas, the transport of natural gas is far more difficult than that of oil. Pipeline transport is the most simple and common method of transporting it, and the gas requires little processing before it can be sent through pipeline networks (in general, removing water and other liquids is the primary concern before transit). Pipelines are a secure and efficient method of transport (Image 5.12), but their high cost and lack of flexibility require time and consumer demand to merit construction. In a new oil field, for example, produced oil can be exported in truck tankers or rail cars until pipeline facilities can be constructed; in natural gas fields, pipeline infrastructure must be constructed before a field can be economically viable.

As the need for pipelines indicates, regional transport of natural gas costs more up front than oil transport, and this basic principle becomes even more pronounced at a global level. Oil, as a commodity, is **fungible**, meaning that it is broadly equivalent in makeup and price worldwide and can be easily transported anywhere. Other fungible commodities like gold and silver also have relatively uniform prices worldwide, primarily because they require no special transport infrastructure. Natural

Image 5.12 Transporting natural gas is more complicated than transporting solid coal and liquid oil. Pipelines are the most efficient means to move natural gas.
Czdast / Shutterstock

gas, while physically relatively uniform, cannot have a true global market because it cannot simply be transported anywhere without the construction of expensive infrastructure and extensive processing. Consequently, the price of natural gas will vary widely, depending on how easy and affordable it is to transport the gas to a specific place.

To be transported overseas, natural gas must be liquefied, converting it into **liquefied natural gas** (LNG). Liquefied natural gas requires a great deal of processing and energy at large terminals before transport, involving compression, refrigeration, and storage, and is therefore much more expensive than conventional natural gas (Image 5.13). Liquefied natural gas terminals and transport ships are also expensive (some of the most expensive infrastructure projects being built now are LNG terminals that cost tens of billions of dollars each), making the transition to a regional LNG economy costly (Image 5.14). The added price of LNG and the resulting difficulty in transporting natural gas between markets means that gas prices vary greatly from region to region. For example, the recent growth in shale gas production in North America has made North American natural gas prices very low, while prices remain much higher in Asia and Europe, where demand is high but local production is low. Islands without their own natural gas production, such as Japan and the Philippines, face the highest prices because only LNG is available. As natural gas production grows worldwide and gas continues to displace coal and oil, LNG use is likely to continue growing, but because of the massive infrastructure investments and higher cost, regional gas production will continue to enjoy a large cost advantage over gas that must be imported from overseas.

Image 5.13 To transport natural gas overseas, liquified natural gas plants and tankers are needed. These use massive amounts of energy to compress and cool natural gas into a liquid for transport via ship.

lastdjedai / Shutterstock

Image 5.14 Liquefied natural gas tanker ships move natural gas across the ocean, but can only deliver to special terminals that regassify the cargo.

Oleksandr Kalinichenko / Shutterstock

VIGNETTE

The Hydraulic Fracturing Controversy

Advances in drilling technology over the past decade have revolutionized the natural gas industry, allowing the exploitation of unconventional resources. New drilling and production technologies, primarily horizontal drilling combined with modern hydraulic fracturing, have allowed economic production from resource reservoirs that were once thought to be impossible to tap because of low permeability, depth, or challenging conditions. Fracturing has been commercially practiced since 1949, but its more visible role in unconventional resource developments since the mid-2000s has fed an ongoing controversy over the safety of the technology versus its role in creating energy security. This controversy has played out in legislatures, courtrooms, and the court of public opinion over the past several years.

In a conventional well, there is sufficient permeability to allow oil and gas to flow or be pumped into the well bore after it is drilled, but in the United States, such reservoirs have mostly been exploited and exhausted. To produce from less permeable rock such as shale, permeability must be artificially created for a well to be able to produce oil or gas. Fracturing accomplishes this by first creating fissures (or fractures) in the target formation and then filling the created spaces with a proppant (usually sand) that keeps the fractures open for hydrocarbons to flow through to the well bore (Image 5.15). The section of a well that is to be fractured is first isolated (so other areas of the well, especially anything in contact with groundwater, will not be fractured), and then large amounts of water with a proppant and a **surfactant** chemical (usually a soap that reduces the surface tension of water and allows it to flow more easily) are pumped into the well at extremely high pressure. This high pressure forces cracks in the rock, which are then filled with the sand in the mixture, which keeps them open after the pressure is reduced. After the fracturing is completed, the used water is cycled back to the surface, where it will be taken for treatment or storage in a depleted rock formation or reinjected for another fracturing operation.

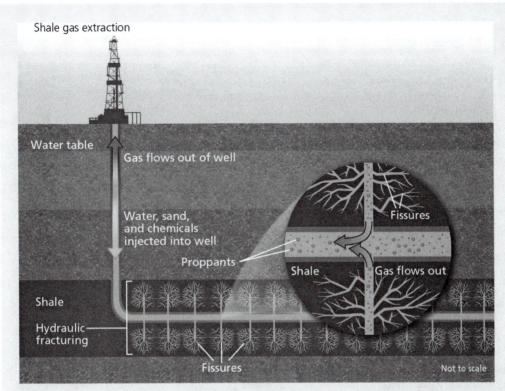

Image 5.15 By pumping proppant, water, and chemicals into the well bore at high pressure, fractures are created and held open with the proppant, increasing the rock's permeability and thus its ability to produce.

The fracturing process has recently attracted controversy for a range of different reasons, some of which are intrinsic to the process and others of which are related not to fracturing, but to oil and gas drilling in general. The original concern, and still the largest topic in this controversy, is the question of whether fracturing contaminates groundwater and, if so, to what extent. Films such as *Gasland* have used anecdotal evidence to suggest that fracturing releases chemicals and produced natural gas directly into the groundwater supply, although this assertion fails to account for several complicating factors. The first suggestion, that fractures penetrate groundwater reservoirs, is almost certainly false in nearly all cases because the fractures created underground cannot penetrate more than tens of feet, let alone the thousands of feet that separate most fractured zones from groundwater. Such reservoirs also have thick impermeable rock layers between them and the groundwater, which ensures that **fracking fluid** would not leak later into the groundwater from the deep reservoirs. In rare cases, the steel casing of wells that have been fractured can leak, leading to the possibility of groundwater pollution, although this can happen in any oil or gas well and is not a direct consequence of fracturing. Clean water regulations in the United States require wells to have both steel and concrete casing around the well bore to minimize the risk of such a leak.

Most incidents of flaming tap water and other natural gas contamination have been demonstrated by geochemists to have been caused by biogenic methane, which is to say natural gas produced by bacteria that

continued

continued

have gotten into groundwater, and not by the deeper thermogenic methane produced by natural gas drilling. A more thorough regulatory framework could potentially help in addressing many of the concerns with water contamination, although industry regulations vary widely between states in the United States and among countries worldwide.

A significant environmental concern with fracturing is how much water it uses. In places such as California with water security issues, the hundreds of thousands to millions of gallons of water that can be used in the fracturing of a well may directly remove water supplies that are needed for agriculture or municipal areas. This concern is geographically specific because many locations can use nonpotable water from deep underground for fracturing and thus do not diminish other water supplies. In regions such as the Middle East, experiments are underway to use seawater for fracking. Still, a balance must be agreed on in many places as to how to equitably distribute water among agriculture, industry, and the population. In especially dry environments, this issue may render fracturing unsustainable.

Other concerns with fracturing have recently been raised, including questions of whether it can cause earthquakes, a condition known as **induced seismicity**. A small number of areas with active fracturing operations have witnessed a large increase in small-scale earthquakes, which demonstrates the need to evaluate producing regions for geographic risk features before large-scale drilling. Geophysical studies have indicated that induced seismicity is not caused directly by fracturing; rather, it is caused by the disposal of large volumes of wastewater into deep rock strata near existing faults that are lubricated, causing more slippage and triggering small earthquakes. Induced seismicity is a particular problem in northern Oklahoma, where an unconventional oil play called the Mississippian Lime produces over 90 percent formation saltwater, which is then disposed of in high-volume wastewater disposal wells in an area of numerous old faults. Because many producing regions lack these faults or large amounts of wastewater disposal, drilling projects must be individually assessed to ensure local safety and come up with acceptable water disposal solutions.

Image 5.16 Specialized equipment is used at each well to prepare the proper mix of water, proppant, and lubricant for a frack.
Bill Cunningham/US Geological Survey/Science Photo Library

Hydraulic fracturing and the associated oil and gas drilling and production are industrial processes involving equipment, noise, and transportation of machinery, which cause concerns for both public safety and the environment, particularly when located near housing areas. Unconventional gas and oil development are responsible for many thousands of new wells, which have led to public and environmental opposition to both fracking and industry development in general. The aforementioned concerns do not represent every objection to the practice, but they do represent many of the main points of individuals opposed to fracking. There are tradeoffs with the practice, particularly with water use and treatment, but most complications with fracturing are geographically specific, rather than indicative of a widespread problem with the practice. These tradeoffs should be considered for every region in which energy exploitation takes place, and local regulators and stakeholders must weigh the benefits against the risks.

Fracturing has played the major role in vastly increasing the supply of natural gas and oil in the United States (Image 5.17). Plentiful natural gas has been replacing coal for electricity generation at a rapid rate, thus reducing greenhouse gas emissions for the first time within the United States, assuming that methane leaks are minimal. As a transitional fuel, the environmental benefits of natural gas produced by fracturing must be weighed against that of dirtier fuels like coal that would replace it in the short term, rather than comparing natural gas to renewables, which are not yet capable of sustaining the majority of our energy supply with current technologies. These many tradeoffs, nuances, and complicating factors have made the controversy over fracturing difficult to balance, as interests across the political spectrum have used it as a proxy issue on environmentalism and business practices in general, with the science often being ignored or distorted.

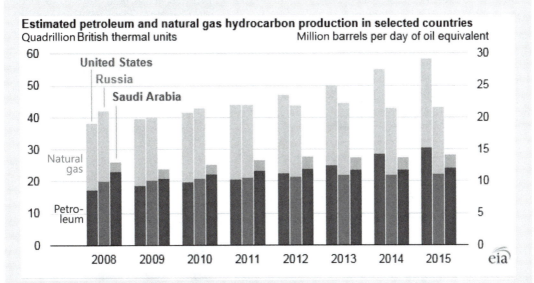

Image 5.17 Fracking has led to a boom in U.S. oil and gas production.

Courtesy of the U.S. Energy Information Administration

Worldwide Natural Gas Production and Consumption

Natural gas occurs in a wide variety of places on earth, with large discovered reserves in many regions and countries. As new unconventional technologies for producing gas have begun to expand production under a greater variety of conditions, global reserves are likely to increase. These increases are likely to be especially pronounced in less developed regions of the world because they have not been explored as thoroughly as the historical oil and gas regions and Western companies attempt to expand their reach to keep production from dropping as a result of depleted reserves elsewhere.

Russia has by far the largest proven natural gas reserves in the world, with a larger total than the fourth through the tenth largest countries combined and more than five times as much the United States (Table 5.2). The rankings of countries by reserves demonstrates the fundamental role of geography in global energy; although Russia has the largest total reserves, many are in remote places such as Siberia or the Arctic Ocean continental shelf, where they are currently too difficult or uneconomic to produce. Some countries that have large reserves may not have the political will or economic ability to produce them, while some countries with much smaller reserves may have sufficiently low production costs or high political will to encourage production. As such, reserve rankings only give a rough picture of how much natural gas certain countries or regions are likely to produce in the future.

The rankings of countries by their natural gas production differ substantially from the rankings of their reserves (Table 5.3). The United States leads the world in natural gas production, largely because of the recent boom in shale gas brought on by

Table 5.2 Proven Natural Gas Reserves by Country, 2016

2016 Energy Information Administration reserves	*Reserves (trillion cubic meters)*
Russia	47.8
Iran	34
Qatar	24.5
United States	8.7
Saudi Arabia	8.5
Turkmenistan	7.6
United Arab Emirates	6.1
Venezuela	5.6
Nigeria	5.1
China	5
World	194.8

Source: Energy Information Administration, "International Energy Statistics," https://www.eia.gov/cfapps/ipdbproject/IEDIndex3.cfm?tid=3&pid=3&aid=6.

Table 5.3 Natural Gas Production by Country, 2016, BP Statistical Review of World Energy 2017

2016 production	Production (billion cubic meters)
United States	749.2
Russia	579.4
Iran	202.4
Qatar	181.2
Canada	152
China	138.4
Norway	116.6
Saudi Arabia	109.4
Algeria	91.3
Australia	91.2
World	3,551.6

improved technologies such as horizontal drilling and hydraulic fracturing, while Russia occupies the second spot by a wide margin. Countries farther down the list with high reserves but low relative production are hindered by the lack of local markets and transportation options. Few countries beyond the United States and Canada have developed their unconventional natural gas potential. Countries far removed from major regional pipeline networks must produce LNG to transport and sell beyond their local markets. Qatar, which is home to the world's largest natural gas field, the North Field, located mostly offshore in the Persian Gulf, is a notable example of a major LNG exporter (Image 5.18). Because of Qatar's lack of nearby consumers, it must rely on LNG exports to support its economy and has consequently become the world leader in the LNG market. Recent projects in Australia have greatly increased its LNG capacity as well, while most other gas-producing countries have yet to invest in further LNG production capacity in recent years.

For both total production and reserves, the worldwide situation for natural gas differs from that of oil. Rather than being dominated by the Middle East, natural gas is spread across many regions (although the Middle East still has large reserves). The high cost to liquefy natural gas means that gas-rich countries without local markets, such as those in the Middle East, cannot export their gas to countries with abundant local reserves, such as the United States. Unlike oil, natural gas pricing varies widely across the globe because of supply, infrastructure, and transport costs (Image 5.19). The U.S. domestic gas market price can vary considerably because of seasonal demand and proximity to production. Eastern U.S. gas prices, once the highest in the country, are now depressed as a result of the huge production boom in Pennsylvania and Ohio. While North American gas prices are now low, often somewhere around $2.25 per thousand cubic feet, or $.08 per cubic meter), gas prices around the world vary widely.

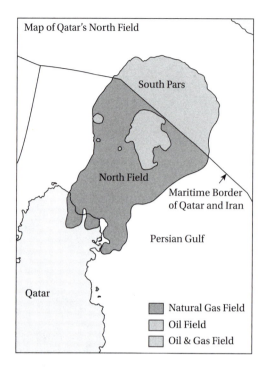

Image 5.18 Qatar's North Field, which crosses the maritime border with Iran in the Persian Gulf, is the world's largest natural gas field.

US Energy Information Administration / Wikimedia Commons / CC-BY-SA-3.0

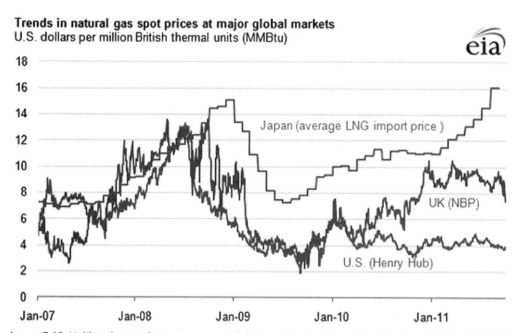

Image 5.19 Unlike oil, natural gas prices vary widely from place to place, with prices especially low in North America and highest in places that must import liquefied natural gas, such as Japan.

Courtesy of the U.S. Energy Information Administration

VIGNETTE

Pipeline Geopolitics in Russia and Europe

The geopolitics of natural gas have been an increasingly important factor in Europe since the collapse of the Soviet Union. Russia has the largest proven reserves of natural gas in the world; most other European countries have limited amounts, often with virtually no production. This has led to Russia being the primary supplier of natural gas to much of central and western Europe, which has become a serious political liability at times because of Russia's erratic relations with the European Union and its former Soviet neighbors. Russian–Ukrainian relations have threatened Europe's supply of gas several times because many of the main Russian pipelines run through Ukraine, making the country vulnerable to politically motivated shutoffs (Image 5.20). The conflict between Russian-backed separatists and the Ukrainian government, coupled with Russia's purported annexation of Crimea, has further exacerbated these tensions and led to more Russian gas transiting via pipelines that avoid Ukraine entirely, although large amounts must still pass through Ukraine. The volatile politics surrounding the gas supply have led to various plans from all sides, with European Union countries and their neighbors considering a variety of ways to decouple their natural gas dependence from Russian influence.

The primary flashpoint for natural gas in Europe has been Ukraine, through which the majority of Russian gas for Europe has been transported. Trade disputes between Ukraine and Russia on gas prices and the cutting of subsidies have resulted in Russia threatening or outright cutting off gas supplies, most notably in 2006 and 2014. The shutoff of gas supplies has been used by Russia as a bargaining tool to place pressure on Ukraine and Europe to pay its set prices, caused shortages and price spikes in several European countries, and threatened a default in Ukraine in 2014 because of arrears on gas payments. Conversely, however, Ukraine's geographic possession of the major gas pipelines to Europe gives it a degree of power over one of Russia's main exports, so Russia has built several pipelines that will allow gas export through the Baltic and Black Seas, bypassing Ukraine entirely (Image 5.21). The 2014 conflict in Ukraine has greatly increased pressure on the European gas market, with many countries in the European Union calling for a move entirely away from Russian gas supplies.

A variety of possible paths could lead Europe away from dependence on Russian gas, but they carry many tradeoffs. A potential solution would be the import of natural gas supplies from other places using LNG. While this

Image 5.20 One of many factors that has led to tension between Russia and Ukraine is the politics of natural gas exports from Russia.

Sergei Chuzavkov / Associated Press

continued

continued

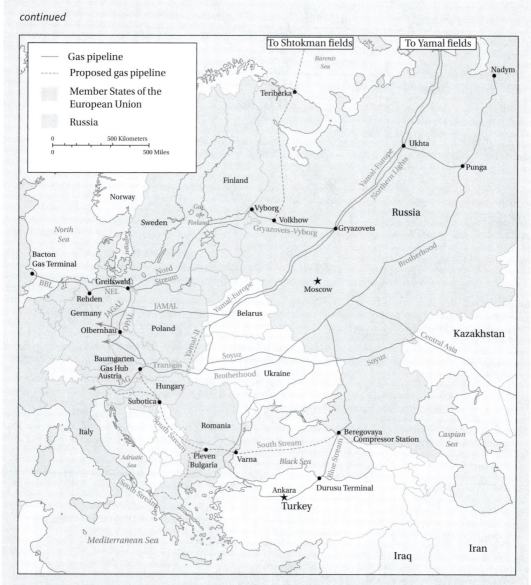

Image 5.21 Natural gas pipelines bring gas from Siberia and the Caspian Sea region to markets in Europe, most passing through Ukraine or Turkey. Courtesy of Samuel Bailey.

would require less infrastructure than some options (because several LNG terminals have already been constructed in Europe), little LNG is produced worldwide because of its much higher price and specific transport requirements. A possibility that has been considered in several European countries is to stimulate their own natural gas production by exploiting newly discovered shale gas reserves. Opposition to the fracking that would be required, as well as a long period before the new reserves could reasonably replace Russian supplies, has caused major political dis-agreements across Europe, with several countries such as France and Germany placing complete moratoriums on production from hydraulic fracturing. Countries such as Poland and Romania have pursued domestic shale gas production, although they have yet to scale production up to the needed levels, and initial drilling has not led to sizeable discoveries of gas. Pipelines under the Mediterranean Sea have also been proposed, which would link Europe to gas production in Israel and Libya, to add to the three pipelines that already supply Spain and Italy with Algerian natural gas. The last possibility for Europe to move away from Russian gas is to transition more heavily toward renewable and alternative energy sources, as has been pursued most vigorously in countries such as Germany. The higher costs, lower output, and inability to go completely renewable without major technological and infra-structural breakthroughs have ensured that renewables cannot significantly impact the demand for natural gas at present, although continued developments and technological advancements could change this in the coming years.

The uneven distribution of natural gas resources across the globe has endowed countries like Qatar with major exportable resources, while other major consumers of natural gas possess almost no resources and are thus highly dependent on imports. Asian countries are the largest consumers of LNG, with Asia accounting for 71 percent of global LNG trade in 2012. Japan is the largest, importing over 40 percent of world LNG alone. Japan's almost complete lack of natural gas resources, combined with the inability to cheaply transport pipeline gas to the islands, makes LNG supplies essential for fueling its economy. Other major importers of LNG in Asia include South Korea, China, and India, while Spain and the United Kingdom are the largest consumers outside of Asia.

Gas Production and Consumption in the United States

Much as U.S. domestic production of oil has increased with the advent of new technologies, natural gas production has grown in a similar although even more dramatic fashion. Although oil production is increasing again, the United States is far from self-sufficient in terms of oil, but the opposite is true for natural gas. The United States has large reserves of natural gas, including dry shale gas, associated gas, and coal-bed methane. These reserves, coupled with new technologies, have continued to increase reserves. The total national gas reserves are sufficient to completely supply local demand for the foreseeable future, leading some to call for natural gas exports, primarily in the form of LNG, to Europe and Asia. In addition, the abundance of natural gas has pushed domestic prices so low that a large amount of coal-based electricity generation is being retired in favor of cleaner burning natural gas generation.

By 2006, conventional natural gas production in the United States had been in decline for several years, and plans to build major LNG import capacity were in place to make up for weak domestic supply (Image 5.22). Unconventional tight gas production technologies have reversed the decline, to the point that many anticipate the United

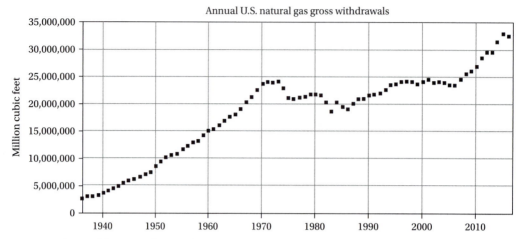

Image 5.22 Natural gas production in the United States has grown dramatically over the decades. Courtesy of the U.S. Energy Information Administration

States to become a net exporter within a decade. This growth has been strong in most places that have been experiencing the tight oil boom, but the eastern United States has been a particularly major center of natural gas growth. The Marcellus Shale, underlying much of Pennsylvania, West Virginia, Ohio, and New York, has become the largest source of natural gas in the United States and is set to continue growing for many more years (Image 5.23). As a result of hydraulic fracturing, the Marcellus has been able to pass 495 million cubic meters per day in 2015, despite producing less than 60 million cubic meters per day in 2009.

Low natural gas prices have had an unexpected consequence in the United States, namely a consistent drop in the use of coal as a fuel for electric generation. Stricter regulations on mercury and sulfur emissions from coal plants have combined with low natural gas prices to drive an increase in natural gas–fired generation at the expense of coal. The U.S. Energy Information Administration estimates that up to 60 GW of installed coal generation capacity will be retired by 2020, primarily because of cheaper and cleaner natural gas competition. Natural gas generation is forecasted to continue to increase for the next several decades, while coal generation is forecasted to remain flat or even decline.

The increases in natural gas production through hydraulic fracturing have been uneven because the controversy over the environmental impact of fracturing has induced several states and municipalities to ban the practice entirely. While many bans have been introduced in regions with no actual natural gas production, the state of New York's 2014 ban will prevent further development of its considerable Marcellus Shale deposits. While drilling activity across the border in Pennsylvania has been increasing steadily since 2010, the inability to fracture wells in New York will cease almost all future production. The extent of economic and environmental impacts or benefits from the ban are unclear, but cross-border comparisons in the coming years will likely reveal the role that the natural gas industry has played in these areas, for good or for ill.

(a)

(b)

Image 5.23 The United States is home to many major shale gas deposits, such as the Marcellus, Eagle Ford, and Bakken deposits.

Image 5.24 New natural gas wells, all using fracking technology, are popping up in many parts of the United States, leading a boom in new natural gas production.
Graeme Shannon / Shutterstock

Methane Hydrates

Unconventional natural gas resources have begun to play an important role in worldwide gas production, especially in North America; other sources of natural gas could have even larger reserves, but also come with greater challenges. **Methane hydrates** are compounds of water and methane, in which the methane is trapped in a crystalline structure that resembles ice, leading it to sometimes be known as *fire ice*. Methane hydrates occur almost entirely in marine sediments off coastlines, and some estimates suggest that there could be more reserves than the entire combined global reserves of oil and natural gas (Image 5.25). Our limited exploration of the continental shelves where these hydrates occur prevents a more accurate gauge of reserves, but they nonetheless possess a large amount of energy that could potentially be extracted and used.

The total amount of methane hydrates worldwide is potentially enormous, but they could be one of the most problematic energy resources to exploit. The technology for developing methane hydrates is still in its infancy, although several countries that lack traditional gas reserves, such as Japan, are attempting to develop their own sources for use in the coming years. Most theoretical methods of extracting methane hydrates will likely be expensive, involving either seabed mining or advanced offshore drilling technologies. If the costs for developing these resources cannot be significantly decreased, they will not likely be able to compete in an open market against shale gas or other natural gas sources that are increasingly being exploited worldwide. The incredible size of expected methane hydrate reserves has continued to stimulate interest, however, and much like other energy technologies in their infancy, hydrates could potentially become a viable energy resource given the appropriate technological breakthroughs or if the economics of other energy resources change drastically.

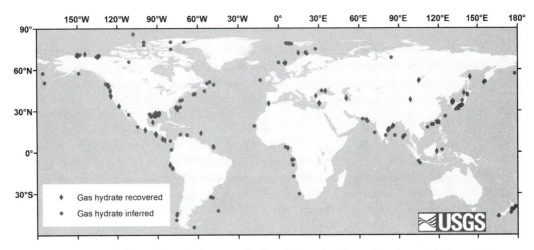

Image 5.25 Methane hydrates can be found along the edges of continental shelves and in permafrost in many parts of the world.

Courtesy of the Woods Hole Coastal and Marine Science Center

Image 5.26 Methane hydrates are trapped in ice and can be burned while still in a frozen state.

U.S. Geological Survey/Science Photo Library

Impacts of Natural Gas Production and Consumption

As a fossil fuel, natural gas is a significant source of pollutants and greenhouse gases, much like oil or coal, yet it is the cleanest fossil fuel in use today. Because natural gas has been steadily replacing coal in power generation and has the possibility of displacing some oil for transport, it is best compared to these fuels to assess its environmental impact, rather than to other energy sources with which it does not directly compete. Given those relationships, natural gas makes for a relatively clean transitional fuel, and the transition to natural gas has allowed the United States to mitigate some of the emissions inherent to our large-scale energy usage. Despite its comparative advantages with coal, natural gas extraction and burning remain significant sources of methane and CO_2 emissions, and natural gas drilling can directly and indirectly disturb local environments.

In direct comparison, natural gas is cleaner than coal in generating the same amount of electricity. Per megawatt-hour, natural gas emits approximately half as much CO_2 as coal, one-third as much nitrous oxide, and one-hundredth as much sulfur dioxide. In a region with adequate pipeline infrastructure, natural gas also saves a great deal of energy through transport because it can be piped directly, rather than traveling on heavy equipment such as trains, as coal does. For other uses such as transit, natural gas offers similar environmental benefits. Compressed natural gas vehicles are growing rapidly in market share worldwide, but have achieved especially significant penetration in Asia. Compressed natural gas power produces fewer pollutants than conventional gasoline engines, and the price of fuel will reflect the more local supply of gas rather than the worldwide oil price. For countries such as Iran and the United States, with large natural gas reserves and growing CNG use, this makes CNG vehicles cheaper to operate, in addition to being cleaner (Image 5.27). Nearly twenty million natural gas vehicles currently operate worldwide, and this transport sector is forecasted to continue growing strongly. Obstacles to CNG use are similar to those faced by other natural gas industries, namely the difficulty

Image 5.27 Compressed natural gas can be used in vehicles and is especially common for use in highly polluted places, such as cities in India.

in storing natural gas because of its density and the difficulty of transporting gas without a robust pipeline network. As natural gas–specific transport infrastructure is constructed, barriers to the adoption of CNG as a fuel will become less significant.

Although it is much cleaner than other fossil fuels, natural gas still emits pollutants and greenhouse gases in combustion, giving it a significant environmental impact. Perhaps the largest concern with natural gas is when methane gas is directly emitted into the atmosphere (either through leaks at wells or pipelines or through indirect combustion). Methane is a much more potent greenhouse gas than CO_2, making its direct emission much more problematic. Per weight, methane absorbs twenty times more heat than CO_2 in the atmosphere, with a lifetime global warming impact between twenty-eight and thirty-six times that of CO_2. Leaks of methane from wells and pipelines could potentially be greatly influencing climate change because of methane's outsized impact on the atmosphere, but the amount of methane emitted is hard to quantify. If significant amounts of methane are leaked, it could potentially be even worse than coal for climate change impact, although coal mining itself is a major source of methane, as well as its many other major environmental impacts (coal mining's methane emissions are also generally impossible to reduce, unlike those of natural gas drilling). Incomplete combustion in power generation facilities can lead to excess methane being emitted, although most modern generators are efficient and do not suffer from this issue. Carbon monoxide is also released during incomplete combustion. When used in home heating, this can lead to the death of persons via carbon monoxide poisoning. The CO_2 emission potential of natural gas is significant in itself, making it more attractive than coal as a transitional fuel, but not ideal as a permanent power generation solution when climate and pollution effects are considered.

In addition to emissions at power generation facilities, natural gas drilling and production can result in significant environmental impacts. The process of hydraulic fracturing has been highly controversial because of its possible problems, although many of these problems are caused by other parts of the drilling process, such as wastewater disposal. Regardless, lax regulations and improper procedures can result in groundwater contamination, induced seismicity, noise pollution, and excessive water usage (covered in the section on fracturing). Further, degradation of well components and valves can lead to direct methane emissions into the atmosphere, and many jurisdictions have yet to create specific regulations to minimize this type of occurrence. The U.S. Environmental Protection Agency has established an agenda to control these types of emissions, and many states are beginning to tighten their regulations in response. If other countries follow this example, methane leakage can be greatly minimized worldwide. While generally cleaner than coal both to produce and to burn, natural gas still can have significant impacts on air, water, and land, especially in the absence of thorough environmental regulations.

Because of its increasing production worldwide and its comparative economic and environmental advantages, natural gas is likely to continue to grow in use worldwide, potentially helping to transition away from conventional fossil fuels until renewable technologies can be developed that can fully replace natural gas. As a transitional fuel, natural gas pairs well with many renewable technologies. Natural gas generation is much more flexible than that of coal or oil, so it can be quickly and easily scaled up and down in a regional grid when the variability of wind and solar causes power levels to drop. Further, unconventional drilling technologies may begin to be widely applied outside the United

States, potentially replicating the U.S. natural gas boom on a global scale. In the case of these developments, natural gas could become an even more integral part of the global energy system, although further advancements will depend on the progress of development of wind, solar, and more experimental renewable energy technologies. For now, natural gas will at least play an increasingly important role in the U.S. energy basket as it continues to replace coal in the transition toward a somewhat cleaner energy economy.

REVIEW QUESTIONS

1. Can all sources of natural gas also contain oil? Why or why not?
2. What are the efficiencies that lead some to suggest natural gas as the transition fuel from fossil fuels to renewables?
3. What other major uses are there for natural gas besides electricity production?
4. Why does natural gas have significantly different prices around the world compared to oil? Would all regions want LNG?
5. What are some of the complications in the fracking debate? Are the problems from fracking uniform or geographically distinct?
6. What are some of the factors that have led to the natural gas boom in the United States?
7. What role do methane hydrates play in the current global energy supply? Why?
8. Why should we be especially concerned about methane releases into the atmosphere?

Image 6.1 Zhangyang13576997233 / Shutterstock

CHAPTER 6

Nuclear Power

The harnessing of atomic energy is one of humankind's greatest achievements; however, it also arouses greater fear, anxiety, and controversy than any other energy resource that we exploit today. Nuclear energy began first with weapons, not electric power plants, but after rapid implementation of atomic power during the Cold War period following World War II, accidents and issues have caused this carbon-free energy source to stagnate despite its obvious advantages with regard to climate change.

The atom was first harnessed to produce nuclear weapons and only later adapted for the generation of electricity. During World War II, both the United States and Germany assembled teams of physicists to develop a weapon that would use nuclear fission or fusion to create a massive energy release and explosion. The U.S. effort, code-named the Manhattan Project, brought together some of the finest scientific minds in the pursuit of this terrifying weapon (Image 6.2). The physicists had to invent both a method for producing highly enriched uranium and plutonium and a means to weaponize these elements to produce a high-energy nuclear explosion. Following the war, it was clear that there would also be peacetime applications for nuclear fission, and work began on creating **nuclear power** plants.

The first successful nuclear chain reactions were achieved in 1942 at the Fermi laboratory in Chicago using uranium and graphite, with cadmium (which absorbs neutrons) to control the speed of the reaction. Small reactors were built during World War II and just after to produce plutonium, and in 1954, the Soviet Union built the first nuclear power plant at Obninsk, which generated 5 MW of electricity and large amounts of heat

Image 6.2 The Manhattan Project developed the first nuclear bombs at Los Alamos National Laboratory in Los Alamos, New Mexico.
Associated Press

used for other purposes. The United Kingdom followed with the first commercial-scale nuclear power plant in 1957, and the first in the United States, the Shippingport Nuclear Power Plant, opened the following year as the first nuclear plant devoted solely to peaceful purposes (Images 6.3 and 6.4).

After these initial successes, nuclear power grew quickly in countries including the United States, the United Kingdom, France, Japan, and the Soviet Union. After a couple decades of rapid expansion, the accidents at Three Mile Island and Chernobyl led to a slowdown in some parts of the world, as well as a complete halt to new nuclear projects in the United States, for several decades. Only now, with growing concern over the impact of fossil-fueled electric generation on the global climate, nuclear power may be moving toward a renaissance, or perhaps not.

As an energy source, nuclear power has always been characterized as the good, the bad, and the ugly. The good: nuclear power produces massive amounts of energy from a small quantity of a resource, and it does so with fewer local environmental impacts than its fossil fuel competitors. The bad: nuclear fission produces unique waste products that are highly toxic and remain so for thousands or even millions of years. The fact that reactors produce fissile products that can also be used for nuclear weapons has become a major proliferation concern. The ugly: although rare, an accident from a nuclear power plant can lead to catastrophic outcomes through the release of radioactive particles into the environment. Despite the rarity of nuclear accidents, the extent of potential damage and the length of time and resources required to clean up the mess have led many to oppose any use of nuclear power. The accident at Fukushima in Japan has once again brought great controversy to the power source (Image 6.5).

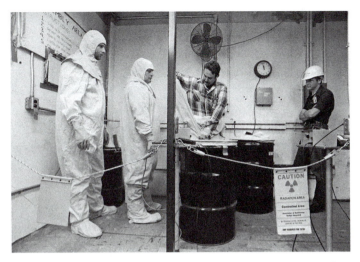

Image 6.3 The Shippingport Nuclear Power Plant was the first such facility in the United States.
Gene Puskar / Associated Press

Image 6.4 The Obinsk nuclear power plant in the Soviet Union was the world's first, built in 1954. Courtesy of RIA Novosti, RIA Novosti archive, image #409173 / Pavel Bykov / CC-BY-SA 3.0.
Pavel Bykov / Wikimedia Commons / RIA Novosti Archive / CC-BY-SA-3.0

Radioactivity and Nuclear Fission

Nuclear power derives its energy from the nuclear forces or strong forces that hold the nuclei of atoms together. Certain larger elements are especially susceptible to being broken apart into more stable elements and can be bombarded by **subatomic particles** in a way that causes their atomic nucleus to break apart. When the nucleus splits, two

Image 6.5 Nuclear power continues to generate widespread skepticism and protest.
Angelika Warmuth / Associated Press

smaller "daughter" atoms are created along with the release of energy and additional particles. When a large enough mass of a fissile material is packed into a small enough space, the particles released from the fission of one atom can be absorbed by the nucleus of other large fissionable atoms, which in turn split, release nuclear energy, and eject subatomic particles that can carry on a chain reaction. This process releases enormous amounts of energy and produces the heat used for nuclear power generation.

All heavy elements, from bismuth onward on the periodic table of elements, spontaneously release particles from their nucleus that we refer to as **radiation**. Radiation comes in three particle forms, each released by different types of radioactive decay: alpha particles, beta particles, and gamma particles. Each has different properties and impacts (Image 6.6):

- **Alpha particles**: an alpha particle consists of a helium nucleus (two protons and two neutrons) and is the least penetrating radioactive particle because it can typically be stopped by something as thin as a sheet of paper and cannot usually pass through animal skin. When an alpha particle is emitted by an atom, it loses two atomic mass units (the mass of a helium nucleus); what remains is the element with an atomic number two lower than the original atom on the periodic table. These particles are the easiest to protect against, but can be harmful when ingested.
- **Beta particles**: a beta particle consists of a single electron or positron emitted by an atom and has more energy than an alpha particle, allowing it to pass unhindered through many common substances. In general, it takes a thin sheet of metal to block a beta particle. Beta particles are formed by a neutron transforming into a proton or vice versa, releasing an excited electron or positron. Because the number of protons determines the identity of an element, this process, known as beta decay, produces an atom of a different element. What remains following beta decay is an atom that has the same mass as the emitting atom but is of the element one atomic number higher or lower on the periodic table.

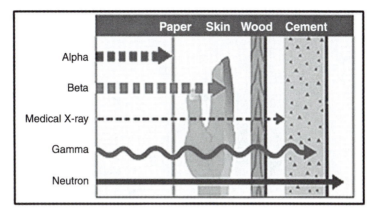

Image 6.6 Different radiation particles can penetrate different substances, with gamma rays and neutrons able to pass through far more substances than alpha and beta particles.
Courtesy of the U.S. Nuclear Regulatory Commission

- **Gamma particles**: a gamma particle is the hardest to detect and the most penetrating radioactive particle; it consists of a high-energy photon emitted by a radioactive alpha or beta particle following radioactive decay. When an alpha or beta particle is released from an atomic nucleus, it may be in an excited state. This excited particle can emit a gamma ray photon to achieve a more stable energy state. Gamma particles, also called gamma radiation, are similar to X-rays and require a large mass, such as lead or thick concrete, to be blocked.

Over time, radioactivity decreases as atoms break down into more stable, and therefore less radioactive, composition. The **half-life** is the amount of time it takes for half of the atoms in any radioactive sample to decay. The faster the rate of decay, the shorter the half-life. For any given element, the half-life is always the same, because half of any sample will decay in a certain given time. The radioactivity of a sample will thus also decrease by one-half in that amount of time. For the long-term impacts of nuclear waste, knowing these rates of decay is essential to properly design waste disposal systems.

Uranium, the base material we use for nuclear power, occurs naturally in three different isotopes or atomic masses (the total number of protons plus neutrons). An element is defined by the number of protons in its nucleus, so different isotopes are distinguished by the number of neutrons their nucleus contains. Natural uranium is primarily U-238, with 0.03 percent composed of U-235 and a tiny proportion of U-234. All three isotopes are radioactive with long half-lives. The half-life of U-238 is 4.5 billion years, while U-235's is 700 million years. These isotopes decay into thorium and continue down over time, eventually settling as a stable isotope of lead. Some of the intermediaries, especially radium and radon, are dangerous naturally occurring radioactive materials. The isotope U-238 is significantly more stable than U-235, to the point that only U-235 can be used to power nuclear power generation. This also explains why much of the earth's U-235 has already decayed over the past few billion years.

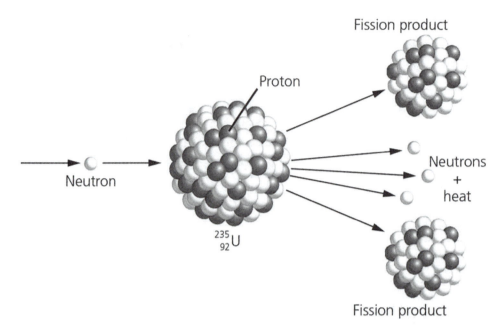

Image 6.7 In a fission reaction, a neutron strikes an atom of uranium 235, producing fission products, neutrons, and heat. The neutrons go on to strike other atoms, creating a chain reaction.

In the early twentieth century, the discovery of free protons and free electrons allowed nuclear scientists to begin experimenting with radioactivity to attempt to manipulate heavy unstable elements. These experiments eventually led to the theory in 1939 that a heavy nucleus absorbing one extra neutron could lead to it becoming unstable and splitting into two lighter elements, releasing all three types of radioactive particles and additional free neutrons. These free neutrons could collide with additional nuclei, leading to a **chain reaction**. At the atomic level, this process releases a tiny amount of energy, about equivalent to one-fifth of the weight of the atom (Image 6.7). However, when you consider that 1 kilogram of U-235 contains 2.56×10^{24} atoms, this means that 1 kilogram of U-235 undergoing fission releases the same amount of energy as burning about 3,000 metric tons of coal. When uncontrolled, such a chain reaction results in a nuclear explosion, the basis for nuclear weapons.

Uranium Mining and the Nuclear Fuel Cycle

Uranium serves as the base fuel for most nuclear power. Unlike other energy resources, except geothermal, nuclear does not derive its energy from the sun. Rather, it relies on radioactive energy from elemental uranium that was created by nuclear fusion within the primordial supernovas that expelled the material that would form our solar system. Unstable, high-atomic-mass elements were present in large proportions in the material that accreted to form planet Earth and have been slowly decaying ever since. Naturally occurring uranium is mostly unsuitable for the generation of nuclear power because of the relative lack of the uranium isotope needed. The majority of naturally occurring uranium, 99.3 percent, is of the isotope 238, commonly abbreviated U-238. Uranium 238 is a stable isotope and therefore is not suitable for nuclear fission. Almost all the remaining 0.7 percent of naturally occurring uranium is isotope 235, or U-235, which is used for nuclear fission.

Uranium is found in the earth's crust in a variety of minerals. Uranium is a relatively common element, forty times more common than silver and five hundred times more common than gold. A class of minerals known as **uraninite** or **pitchblende** (UO_2, U_3O_8, and U_3O_7) is the most commonly occurring and mined uranium ore, while other uranium-bearing minerals include coffinite, davidite, and brannerite (Image 6.8). These minerals are extracted both from underground and from surface mines before being processed into **yellowcake** and transported on for processing through the nuclear fuel cycle (Image 6.9).

To be commercially viable, most suitable sites for uranium mining only need a fraction of the total ore to consist of the uranium-bearing mineral. Concentrations of less than one-tenth of 1 percent up to several percent are sufficient. Since uranium has been decaying for billions of years, other radioactive particles also exist in uranium mines, including radon gas, which is easily inhaled and can cause radiation poisoning, thus requiring significant safety controls at the mine (Image 6.10). Once mined, the ore is crushed and then treated with sulfuric acid to produce yellowcake, which is a concentrated uranium powder, not yet potent enough for use in power plants or weapons (Image 6.11).

Next, the yellowcake must be transported and enriched. The enrichment begins by extracting the uranium from the uranium-bearing mineral by combining yellowcake and fluorine to form uranium hexafluoride, which exists in a gaseous state a little above room temperature. This gas can then be placed into a centrifuge and spun quickly, leading to U-235 concentrating in the center while the heavier U-238 deposits toward the

Image 6.8 Uraninite is the mineral in which much of the world's uranium is found.
Bjoern Wylezich / Shutterstock

Nuclear Fuel Cycle

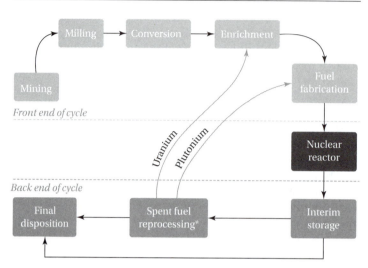

*Spent fuel reprocessing is omitted from the cycle in most countries, including the United States.

Image 6.9 The nuclear fuel cycle begins at the mining phase and moves through enrichment, then consumption, and finally disposal. Open systems, like those in the United States, do not reprocess for repeated use, while closed systems, like those in France, do reuse fuel.
Courtesy of the U.S. Energy Information Administration

Image 6.10 Much of the world's uranium is strip mined.
JuNe74 / Shutterstock

outer edge of the centrifuge (Image 6.12). This process is repeated until the desired concentration of U-235 is enriched in relation to U-238. Earlier processes used gaseous diffusion, wherein the uranium hexafluoride was passed through membranes thousands of times over to separate out the two isotopes.

Once the proportion of U-235 is enriched within the gas, the resulting uranium fuel is then converted from uranium hexafluoride into uranium oxide or uraninite. For power plants, this fuel is usually between 3 and 10 percent U-235, while for a nuclear weapon it is usually enriched to around 90 percent U-235 (Image 6.13). Finally, the enriched uranium is combined with graphite and formed into pellets that are then arranged in fuel rods for the nuclear reactor.

Image 6.11 When uranium is processed, it is first turned into yellowcake before going through enrichment. Courtesy of Patrick Landmann / Science Source.

Patrick Landmann/Science Source

Image 6.12 Uranium is combined with fluorine gas and then processed in centrifuges to separate out lighter U-235, used for fuel and weapons, from heavier and more common U-238, which is not easily fissionable.

U.S. Department of Energy/Science Photo Library

THE POWER OF URANIUM

| 20 grams of **Uranium** | 400 kilograms of **Coal** | 410 litres of **Oil** | 350 cubic metres of **Natural gas** |

Image 6.13 One uranium pellet contains an enormous amount of energy.
Diagram courtesy of the Canadian Nuclear Association

Typically, fuel stays in a nuclear reactor for two to three years and then the reactor is brought offline for refueling. When removed, the spent fuel remains highly radioactive and continues to produce a substantial amount of heat. It is usually stored for one to two years in a pool of water that allows it to cool enough for either disposal or **reprocessing** (Image 6.14). Direct disposal requires longer cooling in tanks to allow the fuel to reach a temperature low enough to permit it to be placed in solid storage devices. This system is employed in the United States.

In many other countries, the fuel is instead reprocessed for further use, which requires that the uranium, plutonium, and other waste must be separated out from the spent fuel. The uranium can be reenriched, capturing any undecayed U-235, which can be used again. Some recaptured plutonium can also be put into mixed fuel with the uranium and reused. All other materials within the spent fuel constitute high-level radioactive waste and contain over 99 percent of all the radiation, necessitating proper

Image 6.14 Fuel rods must be cooled in large pools for years before the fuel can be either reprocessed or disposed of. Courtesy of Patrick Landmann / Science Source.
Patrick Landmann/Science Source

disposal. A system that uses reprocessing is called a **closed fuel cycle**, while one that only uses the uranium once is called an **open fuel cycle**. France, Russia, and the United Kingdom have opted for the closed cycle, while the United States uses an open cycle. Japan was developing a new reprocessing site prior to the Fukushima accident in 2011.

Thermal Fission Reactors and Breeder Reactors

Nuclear power plants work similar to other thermal power plants at their most basic level because they use heat to boil water and produce steam that is then used to turn a turbine and generate electricity. Many different nuclear reactor designs are used to sustain the fission reaction, with the two primary groups being slow neutron **thermal reactors** and so-called **breeder reactors**, which use a faster chain reaction and can use spent nuclear fuel.

Nuclear power plants generate heat energy in the reactor core. Nuclear reactors rely on free neutrons colliding with additional uranium nuclei to sustain a reaction; however, they must do so in a controlled manner. If the reaction speeds up, a nuclear explosion or meltdown can occur, while an overly slowed reaction will fizzle.

Standard reactors hold fuel rods of enriched uranium, usually enriched to no more than 5–10 percent U-235. A moderator substance is then used, which usually identifies the specific type of reactor in use. The **moderator** is the substance used to slow down the free neutrons, thereby ensuring that a chain reaction does not get out of control. A **light-water reactor** uses plain water as its moderator (Image 6.15). A **heavy-water reactor** uses so-called heavy water that contains deuterium (a hydrogen atom that has an additional neutron within its nucleus) in place of normal hydrogen atoms. A small amount of naturally occurring water is heavy, but it must be concentrated from the

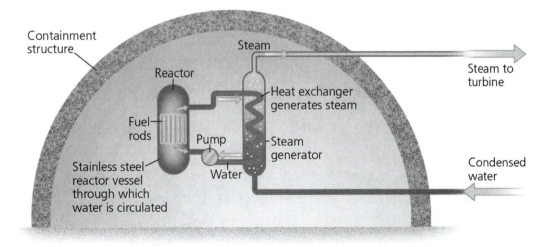

Image 6.15 Light-water reactors are the most common type used for nuclear power. They use normal water as the coolant.

normal water supply to produce heavy water, a laborious process that contributes to low use of this type of reactor today. Some older models of Russian and British reactors use **graphite**, rather than water, as the moderator. The reactor also needs a coolant, which carries heat out of the reactor core for use. In water-moderated reactors, the water also serves as the coolant. Graphite reactors use either water or carbon dioxide gas for this purpose. Finally, the reactor needs a control, which is used to absorb some of the free neutrons and can be modulated to ensure that the chain reaction is sustained but not increased. The control can also be used to stop the nuclear chain reaction whenever the reactor needs to be shut down or serviced. Boron and cadmium are often used as the control substance, and rods of these elements can be inserted and removed from the reactor in the form of **control rods**. Both substances are particularly good at absorbing neutrons.

Other than water-cooled systems and gas systems, both of which use enriched uranium, the **CANDU** (Canadian Deuterium Uranium) reactor uses unenriched uranium. The reactor uses heavy water and a closed-circuit system with smaller fuel bundles compared to standard thermal reactors. These reactors operate at a lower temperature and pressure and can be refueled while in use, making them efficient for operational purposes. The world's largest nuclear power plant, the Bruce Nuclear Generating Station in Ontario, Canada, uses CANDU reactors (Image 6.16). Several nuclear plants outside Canada now also use this technology.

In all types of plants, the reactor core must contain all elements in a manner that can ensure flow of coolant and constant control over the reaction at extremely high temperatures and pressures. The radiation within a reactor core can exceed one trillion times the level a human can tolerate, so the reactor must be well protected. A steel pressure vessel and thick layers of concrete are used to shield the core. The coolants must also be contained when outside the core because of the massive amount of radioactive

Image 6.16 The Bruce Nuclear Generating Station in Ontario, Canada, is the world's largest operating nuclear power plant. Courtesy of Chuck Szmurlo.

material within them. No substances exposed to the core can be released into the environment without creating dangerous radioactive pollution.

Fast neutron reactors, also known as breeder reactors, work differently than conventional nuclear reactors. These reactors do not use slowed-down neutrons and rely on more highly enriched fuel, up to 20 percent U-235. Breeder reactors produce significantly more plutonium than thermal reactors, which means that they create significant amounts of material that can be easily weaponized. Breeder reactors are only used in eight countries—Russia, the United States, the United Kingdom, France, Germany, Japan, India, and China—and were developed as much for the purposes of producing plutonium for weapons as for power generation. Breeder reactors use sodium rather than water or carbon dioxide as their coolant. Because of the weapons concerns associated with these reactors, there are no new breeder reactors being proposed for construction at this time.

Worldwide Nuclear Power Generation

About one-sixth of total electricity generated worldwide comes from nuclear plants. There are over four hundred nuclear reactors in operation in thirty-one countries worldwide. An additional 140 or so naval vessels use nuclear power plants for their propulsion, primarily aircraft carriers and submarines operated by the United States, Russia, the United Kingdom, France, China, and India (Images 6.17 and 6.18). Russia also has several nuclear-powered civilian ships, including icebreakers used for Arctic operations.

Although only thirty-one countries use nuclear power, the source countries for uranium are limited and several do not have nuclear power generation facilities. Foremost among these countries is Australia, which contains the world's largest uranium reserves (Table 6.1). Despite its abundant access to the resource, Australia generates no nuclear

Image 6.17 Nuclear power is also used for transportation, but only for some submarines, aircraft carriers, and icebreakers. This Russian submarine uses nuclear power.
Dmitry Lovetsky / Associated Press

Image 6.18 All of the United States' current aircraft carriers use nuclear power.
Steve Helber / Associated Press

Table 6.1 Top Ten Countries by Uranium Reserves (metric tons), 2009

Australia	1,673,000
Kazakhstan	651,800
Canada	485,300
Russia	480,300
South Africa	295,600
Namibia	284,200
Brazil	278,700
Niger	272,900
United States	207,400
China	171,400
World	5,404,000

Source: International Energy Agency. IEA World Energy Outlook, https://www.iea.org/newsroom/
news/2016/november/world-energy-outlook-2016.html

power and only extracts uranium for export. Countries like Niger and Namibia also only
participate in the nuclear sector through exports of uranium.

Global production of uranium is currently led by Kazakhstan by a large margin
(Table 6.2). The country has the world's second largest reserves, but produces about
two and a half times more uranium than Canada, the world's number two producer.
Other major production is carried out in Africa and other countries of the former Soviet
Union. Uranium production is not subject to as many geopolitical challenges as oil production, but certain centers, such as Niger and Malawi, have experienced major political unrest in recent years.

Uranium is subject to specific safeguards and tough international structures that
control the fuel cycle because of the threat of nuclear weapons proliferation. The

Table 6.2 Top Ten Countries by Uranium Production (metric tons), 2014

Kazakhstan	23,127
Canada	9,134
Australia	5,001
Niger	4,057
Namibia	3,255
Russia	2,990
Uzbekistan	2,400
United States	1,900
China	1,500
Ukraine	962
World	56,252

Source: World Nuclear Association. http://www.world-nuclear.org/information-library/facts-and-figures/uranium-production-figures.aspx

International Atomic Energy Agency (IAEA), a United Nations entity based in Vienna, Austria, is the primary international body that oversees the nuclear fuel cycle and nuclear power plants around the world (Image 6.19). The IAEA is able to inspect nuclear facilities to ensure peaceful use of nuclear power and the safety of nuclear fuel. Only a small number of states that do not utilize nuclear power have not joined the IAEA, and one rogue state, North Korea, has withdrawn from the body.

Image 6.19 The International Atomic Energy Agency is the international organization responsible for enforcing the Nuclear Non-proliferation Treaty and working to ensure the peaceful use of nuclear technology.
Kyodo News / Associated Press

A smaller group of states, the Nuclear Suppliers Group, essentially controls the nuclear fuel cycle, except in a few countries that have developed or are trying to develop nuclear weapons outside the bounds of international law. The Nuclear Suppliers Group consists of forty-eight countries that agree not to sell nuclear technology and fuel to countries outside the regime of the IAEA and the Nuclear Non-proliferation Treaty. Most major components of nuclear power facilities and most fuel enrichment is done by members of the Nuclear Suppliers Group. However, both the United States and the United Kingdom with India and Russia with Iran have signed agreements that circumvent the Nuclear Non-proliferation Treaty.

According to the IAEA, as of the end of 2011, 2,584 TWh of nuclear power was being generated, with 2,087 TWh (or 80.7 percent) in developed countries. In 2012, 437 nuclear reactors were in operation in thirty-one countries (Table 6.3). Despite the dominant position of developed countries in global nuclear generation, over 80 percent of the nuclear power infrastructure under construction today is in developing countries, with China leading the way in new nuclear power plant construction. Three countries are planning to start construction on a first nuclear plant—the United Arab Emirates, Vietnam, and Turkey—and Iran is continuing construction on its first nuclear power plant. The Iranian plant has generated significant controversy because of the likelihood that Iran also has a covert nuclear weapons program related to the project. In 2015, Iran signed an agreement subjecting its nuclear program to significant inspections by the IAEA.

Despite its status as a proven technology, the enormous up-front costs and complex engineering, combined with safety concerns, have caused nuclear power to grow much slower globally than was anticipated in previous decades. Certain countries, such as Japan and Germany, aim to totally abolish nuclear power despite their reliance on enormous fossil fuel imports. This is primarily because of political opposition to nuclear power, especially following the 2011 Fukushima nuclear accident, along with dropping prices for other carbon-neutral power sources. Even fast-growing economies like China and India are expanding their nuclear sectors far

Table 6.3 Top Ten Countries by Nuclear Power Installed Capacity (megawatts), 2017

United States	99,869
France	63,130
Japan	39,752
China	33,384
Russia	26,111
South Korea	22,501
Canada	13,554
Ukraine	13,107
Germany	10,799
United Kingdom	9,918
World	391,744

Source: International Atomic Energy Agency. https://www.iaea.org/PRIS/WorldStatistics/OperationalReactorsByCountry.aspx

slower than coal and other fossil fuel generation. Unless new binding climate treaties necessitate major expansions in nuclear power plant construction, total nuclear power output will likely remain stagnant at around 10 percent of global electricity generation.

Nuclear Power Generation in the United States

As of 2017, ninety-nine nuclear reactors were producing nuclear power in the United States. This accounted for about 19 percent of electricity generation and 8 percent of total U.S. energy consumption, a significant source and the largest non–fossil fuel energy resource for the United States. With this large base, the United States is by far the largest nuclear power generator in the world.

Despite large reserves of uranium, little is mined in the United States. After the Three Mile Island accident, uranium mining in the United States plummeted. In 2014, only about 6 percent of the uranium used in U.S. plants originated in an American mine, while the other 94 percent was imported. The primary sources for this uranium are the former Soviet states of Kazakhstan, Russia, and Uzbekistan, which together accounted for 39 percent of the uranium used in the United States. An additional 20 percent came from Australia, making it the largest single supplier, followed by 18 percent from Canada, 16 percent from major African suppliers (Malawi, Namibia, Niger, and South Africa), and the remaining 2 percent from other suppliers including Ukraine, China, and Brazil.

Nuclear power provides a varying amount of the total grid from state to state, with many parts of the United States using no nuclear power at all (Image 6.20). The most nuclear-reliant state is Illinois, which derives about half its electricity from nuclear power. Arizona, New York, California, Florida, and Texas all use nuclear power for more than 10 percent of their total electricity generation, while parts of the Mountain West, Alaska, and Hawaii use no nuclear power at all.

The number of nuclear reactors in the United States peaked in 1990 at 112. Because some older facilities were decommissioned, only ninety-nine remain in operation today. The last currently operating reactor to begin operation started up in 1996 in Tennessee. Until the Obama administration came to office, no new reactor had been approved since the Three Mile Island accident in 1979. In 2012, two new reactors were approved in Georgia by the Energy Department and twenty-eight additional applications were under consideration by the Nuclear Regulatory Commission at the time of this writing. The process to approve new reactors in the United States is long and cumbersome, which has greatly slowed the potential for growth in the nuclear sector in the United States during recent decades. However, because of increasing public attention to global climate change and growing electricity demand, the possibility for growth over the coming decades remains. Competition from other sectors such as wind and solar is likely to limit future nuclear growth in the United States, however, especially with the drastic reduction in costs for renewables and their comparative lack of political and financial opposition.

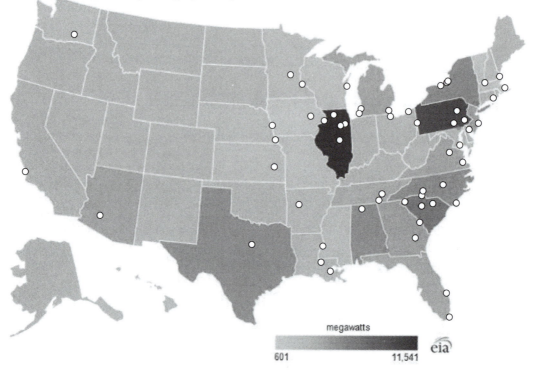

U.S. nuclear capacity (2013)

megawatts

601   11,541

eia

Image 6.20 Nuclear power plants are used across the United States, but more commonly in the eastern states than in the western ones.

Courtesy of the U.S. Energy Information Administration

VIGNETTE

Uranium Mining in Niger

Niger stands out as far different than the other major uranium suppliers of the world such as Kazakhstan, Australia, and Canada. Niger, a landlocked country in West Africa, mostly within the Sahara Desert, is one of the poorest countries in the world, with most people living on less than a dollar per day. Most people in Niger derive their energy needs from the burning of biomass and the small electric grid derives its power from oil, coal, and hydropower—not nuclear power.

Yet, the export economy of Niger is dominated by uranium, which provides well over half of all export revenue (exact figures are not available because of the large, informal economy in Niger). Niger ranks fourth in global uranium production, providing around 7.5 percent of global raw uranium (Image 6.21). How this came to be, and how little it has helped the country, makes Niger an important case study in the energy economy.

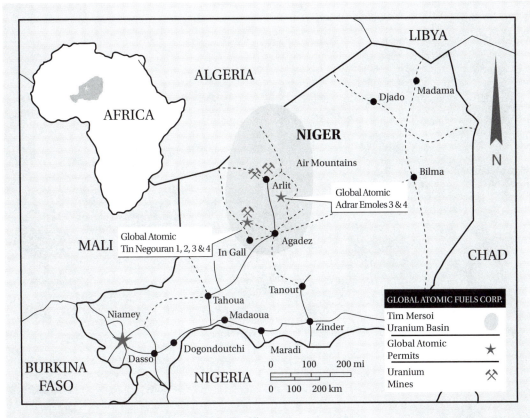

Image 6.21 Map of Tim Mersoi Uranium Basin, Niger

Uranium was first discovered in 1957 in a mountainous part of northern Niger, deep in the Sahara Desert, by French prospectors looking for copper deposits. Commercial mining began in 1971 and grew quickly following the 1973 energy crisis, when France moved to limit its exposure to oil from the Organization of the Petroleum Exporting Countries. The French company Areva invested heavily in mining in Niger, leading to large population growth near the mines in the town of Arlit (Image 6.22).

Yet the benefits of this mining are difficult to quantify. In 2014, the government and Areva entered negotiations on a new ten-year lease of the mines near Arlit. Areva had been paying royalties of only 5 percent and no export taxes to Niger under the prior agreement. As a result, Niger's government was not able to raise significant revenue from the mining, currently accounting for around 5 percent of the national budget, despite uranium being the number one export from Niger, accounting for over half of the country's exports.

Even in Arlit itself, major concerns exist regarding the benefits of mining. Areva is to provide water and power, but electric power is not consistent and neither is water service. Some of these problems may stem from the contracts signed, and others could be tied to systemic corruption in Niger among local officials.

Other recent concerns about uranium production in Niger center on the security situation. Niger's mines are located in a volatile region with significant transborder crime and in an area where terrorist organizations

continued

continued

Image 6.22 The Areva uranium mine in Niger is operated by the French.
Issouf Sanogo / Getty Images

have been active, especially after the civil wars in neighboring Libya and Mali. Although no enrichment occurs in Niger, just yellowcake production, the security of the mines and the uranium ore is a constant challenge.

Most concerning to those concerned with development and social justice is the disconnect between Niger the global energy exporter and Niger the impoverished energy-poor country. Decades of massive energy exports seem to have done little to improve the daily lives or energy security of some of Africa's poorest people.

Meltdowns and Other Nuclear Threats

Because of the dangerous combination of radioactive materials, intense heat, and pressure, nuclear plants have the potential to cause dangerous accidents. Because radiation is invisible and is associated with nuclear weapons, there is great fear surrounding nuclear accidents, which have a tendency to capture the public's attention more than any other energy-related disaster despite being incredibly rare and having a lower total fatality count than many other sources.

If a nuclear plant's containment systems fail, a major release of radiation will ensue. No containment breach has ever occurred during normal nuclear plant operations. Such a breach happens only when the temperature and pressure increase in an uncontrolled manner, leading to a meltdown, or runaway nuclear chain reaction.

One problem that can occur is the result of an uncontrolled reaction. This problem is difficult to achieve in plants with a normal level of enriched uranium because there is not enough material to sustain a major chain reaction. However, some reactor designs, especially certain graphite reactors, yield enough enriched uranium to allow for a runaway chain reaction. In this type of reactor, an error can lead to improper moderation of the reaction with control rods, as was the case in Chernobyl, in which a human error during an experiment resulted in a chain reaction that led to a chemical explosion. The concentration of fuel inside a nuclear power plant is nowhere near the level of enrichment needed to create an atomic blast of the type seen in a nuclear weapon, which requires uranium enrichment of about 90 percent U-235, yet significant contamination of the surrounding environment can occur when a power plant experiences a meltdown.

The more likely problem for most reactors is loss of coolant. Even when a nuclear reaction is stopped entirely, massive amounts of heat continue to be produced by the fuel because different radioactive elements within the fuel assembly continue to break down and produce additional heat. Although this occurs at a much lower rate after the reaction is stopped, residual nuclear decay still produces about one-tenth the amount of heat of a sustained nuclear reaction. The coolant must continue to flow unimpeded during the shutdown and for many weeks following until the fuel is cool enough for removal. The removed fuel must then continue to cool for an additional two years or so before it is ready for transport from the power plant site. Both Three Mile Island and the Fukushima accident were caused by a loss of coolant (Image 6.23). In the latter case, the failure of coolant pumps led to a massive containment breach, while at Three Mile Island the core remained intact.

Image 6.23 Three Mile Island nuclear station became famous when there was a meltdown there in 1979.
Dobresum / Shutterstock

VIGNETTE
The Chernobyl and Fukushima Disasters

Although other small nuclear accidents have occurred, only twice have there been catastrophic failures at a nuclear power plant. The first occurred on April 26, 1986, at the Chernobyl nuclear plant in Ukraine (then part of the Soviet Union) (Image 6.24), and the second started on March 11, 2011, at the Fukushima Daiichi nuclear plant in Japan (Image 6.25). These two incidents galvanized opinions in many parts of the world to oppose nuclear power, beyond those that oppose it for reasons related to nuclear weapons. Each deserves its own attention because they are different from each other and demonstrate why people remain afraid of nuclear power.

The Chernobyl accident was by far the largest nuclear power incident ever in terms of its scale and the death toll that followed (Image 6.26). The accident was caused by human error, when a test being run on an emergency shutdown procedure was improperly performed. It led to a steam explosion and the graphite moderators in the reactor core caught fire. The Chernobyl reactor was not held in a sturdy containment vessel and therefore a plume of radioactive smoke was launched into the atmosphere, spreading nuclear materials over northern Ukraine, southern Belarus (where most of the fallout landed), southwestern Russia, and beyond to other parts of eastern and northern Europe. During the accident itself, thirty-one people died, mainly firefighters and workers at the nuclear plant.

The Chernobyl accident released an enormous amount of radioactive material that led to the evacuation of over 350,000 people from the most affected areas. The city of Pripyat, Ukraine, located near the nuclear plant, was completely abandoned and remains a ghost city (Image 6.27). Although only thirty-one deaths can be

Image 6.24 The Chernobyl nuclear power plant is located in northern Ukraine; however, the meltdown there led to massive contamination in neighboring Belarus and Russia as well.
Peteri / Shutterstock

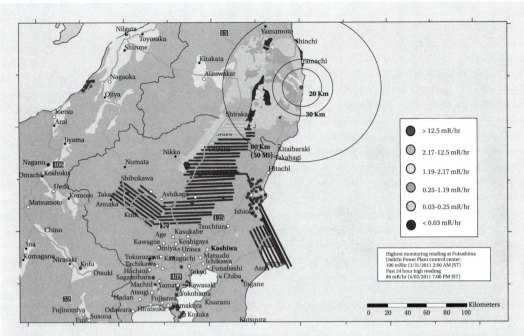

Image 6.25 The Fukushima nuclear power plant is located on Japan's east coast. Much of the fallout from its meltdown spread over the Pacific Ocean, rather than land areas.

National Nuclear Security Administration (NNSA) US Department of Energy / Wikimedia Commons / CC-BY-SA-3.0

Image 6.26 A massive concrete sarcophogus has been built to contain the damaged reactor building at Chernobyl.

Dmitry Birin / Shutterstock

continued

continued

Image 6.27 After the Chernobyl meltdown, the nearby city of Pripyat, Ukraine, was abandoned.

Vodograj / Shutterstock

directly attributed to the accident, the total death toll is likely in the thousands. At least three thousand people have died from cancers, mostly thyroid cancer, linked to the accident, and the Union of Concerned Scientists estimates that up to fifty thousand people could eventually die of cancers caused by the accident.

The more recent Fukushima Daiichi accident was different because it was caused by a natural disaster. An earthquake measuring 9.0 on the Richter scale struck offshore of Japan, triggering a massive tsunami. Only three reactors had been operating at the time because the others were being maintained and refueled, and all three shut down automatically when the quake hit. The subsequent almost 13 meter (43 foot) tsunami breached the 10 meter (33 foot) seawall at the Fukushima plant, flooding the backup diesel generators and other support features at the plant. Several hours later, when backup batteries ran out of power, the coolant pumps failed.

Over the next several days, the nuclear reactors heated up because of the lack of coolant flowing through them, and on March 12 a hydrogen explosion in reactor 1 led to the first release of radioactive materials. Over the next three days, there was a series of further explosions and a plume of radioactive materials was released over the Pacific Ocean and the Fukushima prefecture (Image 6.28).

No one died directly as a result of the Fukushima accident; however, the accident will likely result in increased cancer rates in the area near the plant. Estimates indicate that between 1,500 and 10,000 people will die as a result of the accident over time, while 18,500 died from the earthquake and tsunami. Like Chernobyl, we will never know the exact impact because it is impossible to fully disaggregate cancer cases caused by a nuclear accident to the number of cases that would have happened regardless.

Both Fukushima and Chernobyl raised global concern over nuclear power. Although the accidents differed, they both demonstrate the importance of design safety at nuclear plants to minimize the risk of an accident and especially to minimize impacts if a catastrophic failure occurs. Chernobyl was caused by human error and Fukushima was caused by insufficient backup systems. Although we should not understate the impact of these accidents, in comparison with the number of deaths linked to the production and consumption of fossil fuels, especially coal, the total impact from these accidents is minor.

Image 6.28 The Fukushima nuclear accident was the largest since Chernobyl.
Kyodo News / Associated Press

Nuclear Power and Climate Change

The dangers of nuclear power led to a long-term stagnation in global nuclear generation. However, nuclear remains the most proven carbon-neutral electricity generator, which is bringing increasing attention to it as a source. Since the 1970s, the environmental movement led the global antinuclear crusade. The Green political movement in Europe and later the United States got its start in the antinuclear movement.

Because nuclear power does not produce CO_2, it has attracted renewed attention in the twenty-first century. Although nuclear power cannot displace fossil fuels for most transportation needs, it remains a viable alternative to coal, oil, and natural gas for electricity generation. Some countries committing to reductions in greenhouse gas emissions will likely turn to nuclear as one of the most viable means to reduce their carbon footprint, at least in the medium term, although high capital costs up front could limit overall growth worldwide.

Nuclear Waste

Nuclear waste comes in a variety of forms other than spent fuel. **High-level waste** is responsible for most of the radioactivity in nuclear waste. It is primarily composed of the spent nuclear fuel itself and the leftover materials removed during fuel reprocessing. Much of the waste, other than the uranium and plutonium, is in liquid form at the time of removal and must be cooled for about forty years, at which time it can be made into solid form and mixed with glass. This mixture can then be placed into a storage vessel, along with cooled uranium and plutonium, for long-term disposal.

Intermediate wastes include any substance that has come into contact with the nuclear core. These include the fuel assemblies and the parts of the reactor core that remain after decommissioning of a plant. Additionally, **low-level wastes** are any substances exposed to radiation, either at a nuclear plant or in the medical field. These wastes account for over 90 percent of all nuclear waste produced globally, but only 1 percent of the total radiation produced in nuclear waste. These wastes must be sealed and subjected to proper disposal.

Because of the long-term radioactivity of nuclear waste, disposal options that will keep the waste away from people and the broader ecosystem for hundreds of thousands of years must be considered. This issue has created a quandary because few options are available. Proposed disposal options have included subseabed disposal, a deep hole (greater than 9,000 meters or 30,000 feet), launching the waste into outer space, burying the waste in deep ice sheets (such as in Greenland or Antarctica—and with climate change and the subsequent reduction in the ice sheets, this is a poor idea), island geological burial, deep well injection of liquefied waste, or deep geologic disposal. Most of these solutions carry great risk, such as the potential for a space rocket to explode and release dangerous waste into the atmosphere or for liquefied waste to leak. Deep geologic disposal is the sole option currently considered viable for nuclear waste disposal. Geologic deposits of different types have been considered, including salt beds and stable volcanic formations. A good site must be in a geologically inactive zone, isolated from groundwater, and ideally located far from major population centers. A nuclear waste disposal site must operate with the informed political consent of populations in proximity to the site. The political will to accept nuclear waste remains a large impediment to establishing a proper facility. Geologically, the best location for nuclear waste disposal would be in bedded salt deposits such as those located in western Kansas because the area is geologically stable and the beds are isolated from groundwater. To date, no major geologic disposal site has been opened for waste created in power plants, although the U.S. military operates a site for waste from the nuclear weapons program in New Mexico.

Nuclear waste continues to collect at power plants and other nuclear fuel cycle locations worldwide. In the United States, waste sits at 131 sites in thirty-nine states. Many of these sites are close to major population centers and all require extensive security. Because of the potential for terrorism or accidents, this situation is untenable. However, great controversy remains over proposed final disposal sites, thus leaving the status quo wherein waste is stored all over the country in a variety of sites as the only currently politically viable option. The same problem plays out globally. There are ongoing international discussions to create disposal and burial sites in a few discrete locations that many countries can use, but none has yet gained traction.

In the United States, two sites for waste storage and disposal have been built. One, the Waste Isolation Pilot Plant, near Carlsbad, New Mexico, is operated by the Defense Department and is used to store nuclear waste associated with the nuclear weapons program (Image 6.29). This site is in bedded salt formations, which are stable and dry. However, an accident at the site in 2014 exposed several employees to radiation when a

Image 6.29 The Waste Isolation Pilot Plant facility in New Mexico is the U.S. Department of Defense's nuclear waste repository. The facility is located in salt domes deep underground.
Richard Pipes / The Albuquerque Journal / Associated Press

VIGNETTE
The Yucca Mountain Debate

Since the start of the nuclear age, it has been clear that waste must be disposed of in isolated locations where it can be left alone for hundreds of thousands of years. In 1982, the U.S. Congress passed the Nuclear Waste Policy Act, which called for the creation of a national waste depository that would be selected by 1998. No U.S. state wanted the site and in 1987 Congress designated Yucca Mountain, Nevada (Image 6.30).

Yucca Mountain is located on the edge of the Nevada Test Site, where the United States engaged in various nuclear weapons tests starting in the early 1950s (Image 6.31). This site, although in a mostly uninhabited desert, is only 130 kilometers (80 miles) from Las Vegas, the largest city in Nevada. The proximity to the test site, its location on federal land, being located in an isolated drainage basin, and the fact that it is in one of the driest deserts in the United States all came together to make Yucca Mountain the designated nuclear repository.

The studies to prepare Yucca Mountain as a nuclear repository began following the adoption of Nuclear Waste Policy Act however, there was provision to halt such studies and consider a different location, if warranted. In 2002, Congress recommended the site be designated and preparation on the actual depository began in earnest, with a goal to open the site for initial burial of nuclear waste in 2010.

The site was prepared with tunnels bored into the mountain so that vast chambers could be carved out. These tunnels would then be filled with huge nickel alloy–lined concrete storage containers with everything

continued

continued

Image 6.30 The Yucca Mountain facility in Nevada was built for storing nuclear waste from the nuclear power industry in the United States, but was suspended in 2009.

U.S. Department of Energy/Science Photo Library

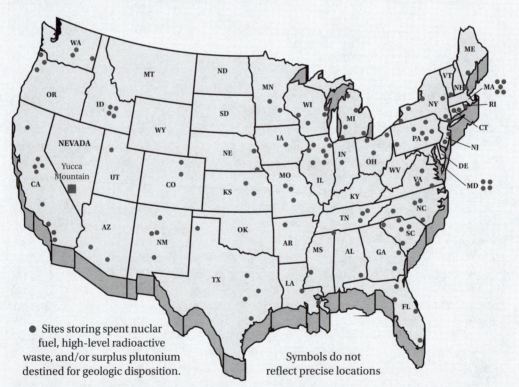

● Sites storing spent nuclar fuel, high-level radioactive waste, and/or surplus plutonium destined for geologic disposition.

Symbols do not reflect precise locations

Image 6.31 Yucca Mountain would allow nuclear waste in the United States to be moved to a single site, rather than the current storage of nuclear waste at individual power plants.

Courtesy of the U.S. Department of Energy, Office of Civilian Radioactive Waste Management from high-level nuclear waste to transuranic waste, both from civilian nuclear power plants and from U.S. military nuclear power and weapons sites.

Construction of the storage site did not end the controversy. Many American politicians, especially those in Nevada, opposed opening the Yucca Mountain site and wanted other locations to be considered. This controversy was elevated to the U.S. presidential election in 2008, when both Democratic candidates—Barack Obama and Hillary Clinton—opposed the site as part of their campaign to win the Nevada caucuses early in the nominating process. The Democratic leader in the Senate, Harry Reid of Nevada, also led opposition to the site.

In 2010, President Obama instructed the Department of Energy to end consideration of the site, just as enough was completed for it to soon open. In 2010, the department filed with the Nuclear Regulatory Commission to withdraw the Yucca Mountain site from consideration and close down the project. The decision went to the federal courts and 2014, U.S. nuclear power plants were allowed to suspend payments to the nuclear waste recovery fund, which was to fund the national nuclear repository and had already expended billions of dollars on the preparation of Yucca Mountain.

Many U.S. states and cities that are home to temporary nuclear storage locations have filed suit to reconsider Yucca Mountain and to begin storage of thousands of metric tons of nuclear waste at the site. Currently, the waste continues to languish at nuclear power plants and defense facilities, which is far less safe than geologic storage in an isolated location. Whether these parties win or another site is chosen, eventually the waste will need to go somewhere other than where it sits now.

container broke. The second site, Yucca Mountain in Nevada, was designated by Congress as the repository for nuclear waste from civilian plants. The site had neared completion in 2009, but was suspended by the Obama administration.

Cost of Nuclear versus Other Energy

Nuclear is expensive compared to other energy sources, and the long-term costs of spent fuel storage and disposal plus the costs of a major accident, which may be borne significantly by the government, are not easily estimated.

Although advanced nuclear cost is estimated to be dropping, the cost of wind and solar has been plummeting at a much faster rate. Unless new technologies and cost savings plus perhaps a carbon tax are applied, nuclear is at a price disadvantage compared to many other sources of energy; thus, little interest in expanding nuclear power exists, especially in the United States.

REVIEW QUESTIONS

1. Why can we not use U-238 to generate electric power?
2. Nuclear power developed as a byproduct of weapons research. Why did it first develop in the United States, the Soviet Union, and the United Kingdom?

3. What are the basic differences between alpha, beta, and gamma particles? Why do these differences matter when designing a nuclear power plant and storing nuclear waste?

4. Why is so little of the uranium in the earth's crust U-235? Has this always been the case?

5. What is the difference between an open and a closed fuel cycle? Which one does the United States use? France?

6. Why does graphite make such a good moderator?

7. What is the one type of nuclear reactor that does not need enriched uranium and how does it work without it?

8. Why are global nuclear watchdogs mainly interested in uranium enrichment above 20 percent?

9. Which countries are major uranium suppliers, but do not use any uranium domestically for power generation?

10. Why should Three Mile Island not be considered a disaster in the same way as Chernobyl or Fukushima?

11. What is the only nuclear repository currently in use? What is the fate of its larger cousin, currently suspended?

12. Is climate change likely to lead to a nuclear renaissance, or are there too many obstacles? Which obstacles do you think will most likely keep nuclear power from making a comeback?

Image 7.1 Lafoto / Shutterstock

CHAPTER 7

Hydropower

Humans have harnessed power from moving water for thousands of years. In ancient times, water wheels were used to power saws to cut wood and later to grind grain. By the Middle Ages, many industries based themselves near flowing water to turn paddle wheels (Image 7.2). These wheels powered mills, which were able to use the mechanical energy for many small-scale industrial processes.

In the United States, early industrial production centered along the Fall Line, a region where eastern rivers flow downstream through many waterfalls. Early industrial production of products, including the spinning of yarn and weaving of fabric, concentrated in New England communities along the Fall Line. Major armories, including the Harpers Ferry gunworks, also were built along fast-flowing rivers (Image 7.3).

In 1827, the first submerged hydro turbine was built in France. A few years later, the first turbines utilizing dams and penstocks were built in Germany. **Hydropower** was not used to produce electricity in the United States until 1880, when lamps were powered at the Wolverine Chair Factory in Grand Rapids, Michigan. Two years later, the first hydroelectric power plant was built on the Fox River in Wisconsin.

Today, hydroelectric dams have been built across the world and are the largest source of renewable electricity in the global energy system, accounting for a significant portion of global electric power production. While earlier use of water power also accounted for direct mechanical energy, almost all power derived today from flowing water is used to produce electricity.

Image 7.2 Water power was used long before the development of hydroelectricity in mills and other small-scale industrial uses.

Inavanhateren / Shutterstock

Image 7.3 This water wheel powered a grain mill in the nineteenth century.

Larry Knupp / Shutterstock

The Water Cycle

The harnessing of hydropower is reliant on the earth's water cycle to harness its energy (Image 7.4). This energy is derived from solar energy, like most of our other energy resources. Solar energy both evaporates liquid water from the oceans, rivers, and lakes

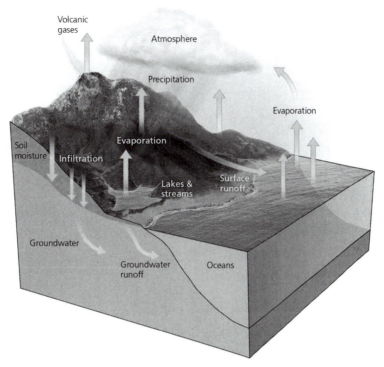

Image 7.4 Solar power indirectly powers hydropower, because solar energy drives the water cycle, which leads to precipitation and the downhill flow of water.

and sublimates ice, which then becomes atmospheric water vapor. When that water vapor precipitates as rain and snow back to the earth's land surfaces, it enters the various drainage basins in river systems. The energy that is released as that water flows downstream toward the oceans can be harnessed through a wide variety of hydro systems.

Because this system works on a continuous cycle, hydropower is a renewable resource. However, it is not a resource of continuous sustained capacity, because there is both seasonal and annual variance in precipitation. Precipitation falls in far-from-uniform fashion, with dramatic differences from place to place. The world's wettest place is Cherrapunji, India, on the Himalayan slopes, with 11,870 millimeters (467 inches) of rain on average, while the driest is in parts of Chile's Atacama Desert that receive about 1 millimeter (0.04 inches) of rain per year. Within the United States alone, the differences range from an average of 11,684 millimeters (460 inches) at Mount Wai'ale'ale, Hawaii (Image 7.5) to only 60 millimeters (2.4 inches) on average per year in Death Valley, California (Image 7.6).

Beyond total rainfall, hydropower requires precipitation to then flow into riverine systems, which are also not uniformly distributed. Some wet areas have little hydropower potential because of the nature of hydrologic features in the area and topography. Other areas that are arid may have large hydropower potential because rivers

Image 7.5 The island of Kauai in Hawaii has the highest rates of precipitation in the United States.
MNStudio / Shutterstock

Image 7.6 Death Valley, California, has the lowest rates of precipitation in the United States.

with particularly good features for harnessing hydropower run through them. Some of the world's most famous hydroelectric projects, like the Hoover and Aswan Dams, are in deserts (Image 7.7).

Hydropower derives its potential from the flow of a river. The steeper the topography, the greater the flow speed. The larger the river, the greater the volume. A combination of these factors leads to higher kinetic energy where there is large water flow moving at higher speeds. Siting for a hydropower project requires more than simply high speed and flow; it also requires geology that will allow dam construction and geology behind the dam that will allow for the creation of a large **reservoir**, or manmade lake (Image 7.8). This reservoir allows for the storage of water so that hydropower can be harnessed

Image 7.7 The Hoover Dam on the Colorado River is not the largest, but is probably the most famous hydroelectric dam in the United States.
Andrew Zarivny / Shutterstock

Image 7.8 The reservoir behind a dam stores water and produces significant pressure to drive the turbines in the hydroelectric plant.
Shutterlk / Shutterstock

when there is demand, rather than merely at times of peak flow. Geologic conditions in which a river flows through a canyon or passes through a narrow gap provide the ideal location for a hydropower project because they create a narrow space in which to build a dam. Additionally, the geology must be stable so the ground around and under a dam will not give way and break from the enormous pressure that builds up behind the dam. Because of the unique factors required, the sites appropriate for large hydropower projects remain limited.

Hydroelectric Dams

All large-scale hydropower projects are built with a **dam** and an accompanying reservoir (Image 7.9). The dam is constructed to block the river and allow people to control the flow of water below the dam. Water used for the production of electricity is allowed to flow through the dam by opening gates at the intake point on the reservoir, or upstream, side of the dam. The water then flows through the **penstock** and into a turbine, where it spins the turbine blades, just like steam or gas does in conventional thermal power plants (Image 7.10). The turbine then turns electromagnets in the generator, creating electricity. The water continues beyond the turbines and then flows out of the dam, often at high pressures, enters the river below, and continues to flow downstream.

Large dams must be able to bear enormous amounts of pressure and varying amounts of water, including the occasional massive influx from floods or heavy rains. Dams must therefore be able to handle water in ways other than draining through the penstocks and the hydroelectric plant. During major water influxes, a dam will also make use of a spillway, which, like the penstocks, is sealed with a gate when not in use. Spillways are in place both to allow water to flow when hydropower is not needed and to handle excess flow during high-water events. This is necessary to avoid **overtopping**, where a reservoir level rises above the top of the dam and causes water to flow uncontrolled over the dam and into the river course below, which can cause damage to the dam itself and lead to downstream erosion as well.

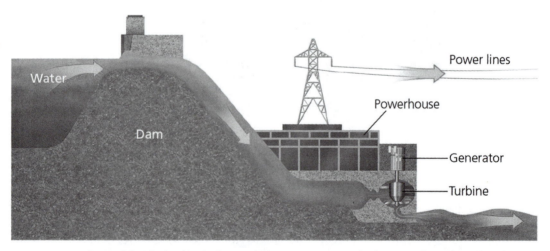

Image 7.9 A hydroelectric dam works by opening penstocks that water flows through and turns a turbine. That turbine drives a generator, like a steam-powered plant.

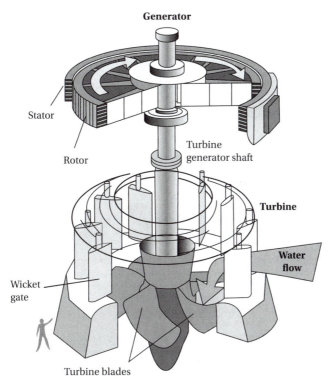

Image 7.10 The hydropowered turbine turns through the flow of running water, rather than steam, like in thermal power plants.

Courtesy of the U.S. Department of the Interior, USGS Water Science School

Another challenge that many large-scale dams will face, especially those in arid areas where river basins are seeing lower flow than in the past, is a lowering of reservoir levels below the spillway and penstock gates. If a reservoir drops below these entry points on the dam, a reservoir enters a state called **dead pool**. At this point, the dam is no longer able to produce electricity and there is no more water flow to the river below the dam. This can become a particular problem after extended droughts or excessive water use upriver from dams, so dammed rivers require careful water resource management to truly be classified as a renewable resource.

Hydroelectric power stations are very efficient compared to thermal electric production systems. Between 70 and 90 percent of kinetic energy is converted to electric energy in a hydroelectric system, with the exact efficiency determined by the design of the penstock and turbine. Hydroelectric dams also provide an ability to quickly ramp up or down the total production of electricity because more or less water can be sent through the system to the turbines, whereas thermal systems can take hours to change because water must be heated in boilers to increase power output.

Image 7.11 The Chief Joseph Dam along the Columbia River is a run-of-the-river dam, rather than using a reservoir. It is one of the largest such hydroelectric projects in the world.
Rita Robinson / Shutterstock

Small-Scale Hydropower

Many major rivers that have hydroelectric potential have already been tapped, especially in the developed world. However, new designs are vastly changing smaller scale hydro projects. Early hydro projects that made use of rapidly flowing water to power mills were commonplace prior to the development of large dams; however, fewer small projects were implemented in the first half of the twentieth century, when most of the large dams in developed countries were built.

Small-scale systems are also known as **run-of-the-river systems** because they do not rely on the damming of a river or stream. Instead, these systems insert turbines into flowing water directly. Such systems run variably, depending on the speed and volume of water in a river at a given time, so they are not responsive to the needs of the power grid, unlike large-scale dams. A second form of run-of-the-river systems diverts water from a flowing river into pipes that then run the water through turbines prior to reentering the river at a lower elevation (Image 7.12). These systems can vary in power produced by modulating how much water is diverted from the river course for use in the power plant, but are still unable to store water for later use because they lack a reservoir.

Small-scale systems are ideal for isolated communities that would otherwise be off the power grid and for environmentally sensitive locations where a dam and reservoir could not be located (Image 7.13). So-called micro hydro systems can even produce small amounts of power for a single family or small village.

Another type of hydro system used for the storage of energy for later use is called a **pumped-storage** hydroelectric system (Image 7.14). In such a system, excess power produced via other means, whether renewable or nonrenewable, can be expended in pumping water to a reservoir on high ground. When the power is needed later, the

Image 7.12 The Machu Picchu run-of-the-river power plant diverts water through a tunnel, reconnecting with the Urubamba River at a lower elevation downstream, powering turbines as it flows through the tunnel.

Image 7.13 Small-scale hydro projects can provide enough power for single facilities or small communities.

Imfoto / Shutterstock

water can be released and turned back into electricity as it flows through turbines on its way back down to the lower elevation reservoir. Pumped-storage systems have been suggested as a solution to the variability of renewable energy technologies like solar and wind, and some countries have begun experimentation with them for this purpose.

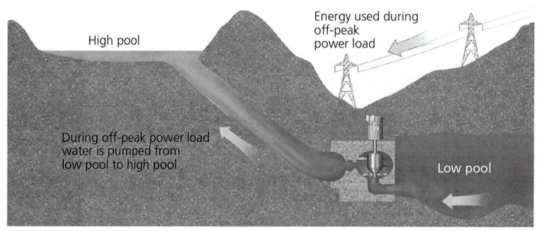

Pumping cycle

Energy used during
off-peak
power load

High pool

During off-peak power load
water is pumped from
low pool to high pool

Low pool

Off-peak power load

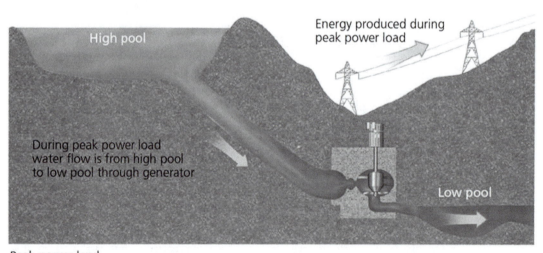

Generating cycle

Energy produced during
peak power load

High pool

During peak power load
water flow is from high pool
to low pool through generator

Low pool

Peak power load

Image 7.14 Pumped-storage systems use excess power to pump water to a reservoir up a hill, which can then be released and flow through hydroelectric turbines to provide electricity when demand is higher.

Worldwide Hydropower Production

Hydropower accounts for the largest share of global renewable electricity production and is by far the most established renewable technology in the global energy basket. Over 80 percent of global renewable electricity is produced by hydroelectric projects. The resource also continues to grow in use because many developing countries continue to build large hydroelectric projects (Table 7.1). Similar growth has not been seen in developed countries, where most viable hydroelectric project sites have already been developed.

In 2012, 13.4 percent of the electricity in developed countries was produced by hydroelectric plants. Hydropower in the developed world is already highly developed and there are few additional sites for expansion (Image 7.15). Developing countries

Table 7.1 Top Ten Hydroelectric-Generating Countries for 2011 (in terawatt-hours)

China	698.95
Brazil	428.33
Canada	375.8
United States	344.68
Russia	167.61
India	130.67
Norway	122.08
Japan	91.71
Venezuela	83.67
Sweden	66.56
World	3565.45

Source: International Energy Agency. Annual Report, available at https://www.iea.org/newsroom/news/2016/november/world-energy-outlook-2016.html

Image 7.15 The Itaipu Dam, shared by Paraguay and Brazil, is the world's second largest hydroelectric dam.

Stefano Ember / Shutterstock

have significant room for growth in hydropower, which by 2011 was already providing 18.6 percent of the total electric generation. Although China produces the most hydropower in total, the country that is most reliant on **hydroelectricity** is Norway, which produces over 99 percent of its electricity from hydroelectric sources (Table 7.2).

Table 7.2 World's Ten Largest Hydroelectric Dams by Installed Capacity (in megawatts)

Three Gorges Dam	China	22,500
Itaipu	Brazil/Paraguay	14,000
Guri	Venezuela	8,850
Tucurui	Brazil	8,370
Grand Coulee	United States	6,809
Longtan	China	6,426
Krasnoyarsk	Russia	6,000
Robert-Bourassa	Canada	5,616
Churchill Falls	Canada	5,428
Bratsk	Russia	4,500

VIGNETTE
China's Three Gorges Dam

One of the biggest symbols of both the promise and the pitfalls of hydroelectric power can be found by examining the Three Gorges Dam in China (Image 7.16). The Three Gorges project is the largest hydroelectric project ever built and the largest dam in the world. The steel and concrete dam is 2,335 meters long (7,661 feet) and 181 meters tall (594 feet), spanning across the Chiang Jiang (Yangtze) River, the longest river in China and, indeed, all of Asia (Image 7.17). The dam houses an enormous electric generation facility with thirty-two turbines rated to 22,500 MW, greater than any other hydropower station

Image 7.16 China's Three Gorges Dam is the world's largest hydroelectric dam.

Thomas Barrat / Shutterstock

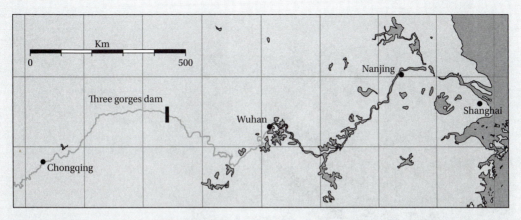

Image 7.17 The Three Gorges Dam lies upstream of major cities along the Chiang Jiang (Yangtze) River. User:Rolfmueller / Wikimedia Commons / CC-BY-SA-3.0

ever built. The dam not only brings cleaner power to parts of central China, but also promises to help control floods and bring economic development. However, the project also brought great controversy that helps demonstrate the potential pitfalls of a massive dam.

On the positive side, the Three Gorges Dam has accomplished much that it promised. In 2014, it produced more electricity than any other hydroelectric dam in the world, overtaking the Itaipu Dam on the Paraguay/Brazil border as the largest producer. It has also helped shipping along the Chiang Jiang River by steadying the river's flow between the wet and dry seasons. The dam is also likely to mitigate future floods along the river, which is notorious for its occasional catastrophic flooding that has killed thousands.

On the negative side, however, the Three Gorges Dam has caused significant problems to both people and the environment. The costs to those living upstream of the dam were felt first because the dam's construction led to the displacement of almost 1.3 million people, according to the government. Widespread allegations exist that much of this resettlement was forced in a manner where people could not choose where to move and that large amounts of money set aside for resettlement were redirected to other uses by corrupt officials. In addition, the creation of the massive reservoir behind the dam not only flooded cities and towns, but also destroyed many archeological and natural sites that the Three Gorges region was famous for.

The impacts to the natural world may be even greater from the dam. The baiji, or Yangtze River dolphin, one of only two freshwater dolphin species, is now extinct (Image 7.18), with the Three Gorges Dam pointed to by many as the final straw leading to its extinction. Many other endangered aquatic species are also at risk. Downstream from the dam, the loss of sediment in the river may exacerbate flooding in many areas and pose a risk to Shanghai, China's largest city, located at the mouth of the Chiang Jiang River (Image 7.19). Shanghai is built on sediments that must be replenished, and with sea-level rise and the dam upstream, the city faces additional threats. Upstream, the reservoir will see increases in sedimentation that will eventually render the hydropower project useless (estimates vary from fifty to one hundred years before this happens). Landslides are also becoming common along the reservoir.

The greatest risks related to the dam come from the potential for it to fail because the river valley beyond the dam toward Shanghai houses hundreds of millions of people. The dam sits along a fault

continued

continued

Image 7.18 The development at Three Gorges has impacted many species and is a contributor to the extinction of the Baiji River dolphin.

Chinatopix / Associated Press

Image 7.19 Massive growth in energy demand in China and pollution from coal-fired power plants helped push for the Three Gorges project.

Chuyuss / Shutterstock

line, and a catastrophic earthquake could lead to dam failure and the release of the reservoir water. In addition, the dam is a major strategic issue in that an air attack on could blow up the dam, resulting in the same catastrophe. Some people in Taiwan have threatened that exact course of action, were China to invade.

It is impossible to render a positive or negative verdict on massive hydro projects like the Three Gorges Dam. The view depends on the individual and where priorities lie between assets such as cleaner electricity and flood control on the one hand and loss of homes and environmental damage on the other.

Hydropower Production in the United States

In 2012, the United States derived 7 percent of its total electricity from hydropower, which also accounted for 56 percent of renewable electrical generation. Hydroelectric generation in the United States is concentrated in the western states, with four of the top five generating states located in the region. The top producing state is Washington, which produces 29 percent of all hydroelectric-derived electricity in the United States, including that from the single most productive U.S. facility, the Grand Coulee Dam (Image 7.20). Rounding out the top five are Oregon, California, New York, and Idaho (Table 7.3).

The largest number of large hydroelectric dams is concentrated in the Columbia River system in the Pacific Northwest region (Image 7.21). About half of these dams were built and are operated by the U.S. Army Corps of Engineers. The earlier dams built in the Tennessee River system are also operated and built by the U.S. government, under the auspices of the Tennessee Valley Authority, a project of the New Deal launched during the Great Depression as a means to provide jobs

Image 7.20 The Grand Coulee Dam in Washington state is the largest hydroelectric dam in the United States.

Edmund Lowe / Shutterstock

Table 7.3 Five Largest Dams in the United States by Installed Capacity (in megawatts)

Grand Coulee	Washington	6,809
Chief Joseph	Washington	2,620
Niagara Falls	New York	2,525
John Day	Oregon/Washington	2,160
Hoover	Arizona/Nevada	2,080

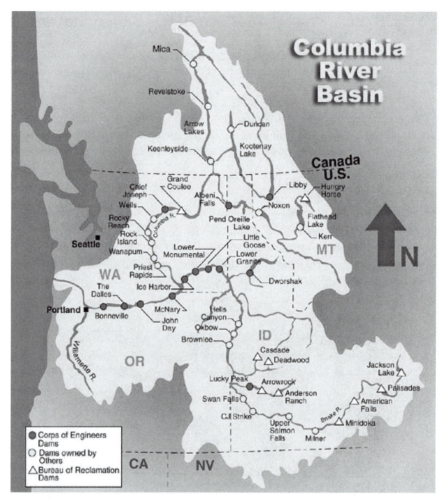

Image 7.21 The Columbia River system in the Pacific Northwest region is home to the most hydrolectric dams in the United States.

Courtesy of the U.S. Army Corps of Engineers

and to bring electricity to poor areas of the American southeast. A third series of major government dams was built along the Colorado River, including the Hoover and Glen Canyon Dams. This large involvement of the government contrasts starkly with most other electric power production in the United States, which tends to be built and operated by private companies or local government utilities, rather than the federal government.

Although many dams are spread around the United States, most do not produce hydroelectric power; rather, they were built solely for water storage and flood-control purposes. For the most part, dams in these regions are not suitable for electricity production.

Environmental Considerations

A series of environmental issues must be considered for all dam construction, including hydroelectric projects. While some concerns can be mitigated, others are always a concern in the construction of dams.

One important consideration is **siltification**. Rivers carry vast amounts of silt, dissolved sediments that travel from the mountains down to a delta. Silts cause rivers to become murky and able to support many species of fish and plants that cannot survive in clear waters. Silts are also deposited on river floodplains, providing important nutrients for the soil. When a dam is built, water flow halts in the reservoir, leading to the silts settling out of the flowing water and onto the bottom of the reservoir. In rivers with high silt content, over time the reservoir fills with silt, which lessens the hydroelectric potential and eventually builds up silt all the way to the dam penstocks, rendering the hydroelectric dam useless. A reservoir can be dredged to remove some of the silt, but this process is expensive and only mitigates the problem temporarily. Some massive projects in especially silty waters, such as the Three Gorges Dam in China, are estimated to have a total life span of well under one hundred years because of siltification. After this period, these particular dams will have to be decommissioned and removed.

Dams can have major environmental impacts upstream from the dam and reservoir. One of the most notable impacts is the interruption dams have on the migration of aquatic animals. Smaller fish that migrate, including salmon, travel to upstream locations to breed, although they spend their lives in the ocean. Salmon always spawn in the same creek where they were born, but a dam stops them from being able to return. This problem can be mitigated by **fish ladders**, a series of steps with water flowing over them that allow fish to jump up the side of a dam and continue on their migration (Image 7.22). The Columbia River watershed in the United States and Canada is an important salmon source area, and the huge hydroelectric projects there had to be retrofitted with fish ladders to mitigate their impact on this fishery. Larger aquatic animals, such as Amazon and Yangtze River dolphins, have been impacted as well (Image 7.23), and unlike fish, they are not able to utilize ladders or other mitigation efforts to allow them to travel. In addition to its other impacts, the Three Gorges Dam may have led to the total extinction of the Yangtze dolphin.

Dams have a variety of impacts downstream from the dams themselves. During extended droughts or water flow disruptions, the water level in reservoirs can drop enough to cause dead pool, when the water level drops below the point at which it can flow past the dam. This causes the river on the downstream side to dry up partially or completely, leading to major disruptions for humans or animals that rely on the river's water supply. Downstream, the dam also impacts water quality and temperature. A lack of silt and the nutrients that come with it leads to water that cannot support a high level of plant and fish life. Also, downstream water is often significantly colder because the water flowing through the dam comes from deep within the reservoir, not from the surface. This is necessary to maintain the water pressure on the turbines needed to maximize electric output. Dams located in hotter climates, such as deserts,

Image 7.22 Fish ladders provide a way for migratory fish to travel upstream where dams block their progress.

Rigucci / Shutterstock

Image 7.23 River dolphins are particularly at risk when dams are built because they are not able to travel beyond a dam. The growth of dams along the Amazon is threatening its river dolphins, like on the Chiang Jiang River in China.

Anirut Krisanakul / Shutterstock

Image 7.24 Salinization of soils happens in arid areas where dams control flooding that would otherwise wash away these salts. Courtesy Kaj R. Svensson / Science Source.

Kaj R. Svensson / Science Source

often feature rivers with a relatively high temperature and therefore species adapted to such temperatures. Downstream from a dam, this temperature change can harm both aquatic plants and animals, as has been seen in the Colorado River system in the United States.

Dams also regulate downstream flooding. This regulation may be beneficial to human populations living along a river, but it has massive ecological impacts. Not only is silt not deposited, but also the natural rhythm of a river valley is halted, which causes changes in how a river flows, impacting the wildlife living in the watershed. In deserts, this regulation can also lead to soil **salinization**, in which salts located deep in a soil profile leach to the surface (Image 7.24). When floods occur, these salts wash away, but without such floods, the salts remain and damage soils. Most food crops are salt intolerant and cannot grow under such conditions.

Impacts on People and Society

Not only does the construction and operation of hydroelectric projects impact natural cycles, but also it can also cause havoc for human populations. Foremost among hydroelectric concerns is the possibility of a dam failure. Hydroelectric dams hold back billions to trillions of liters of water in a reservoir and must be designed to handle many different risk factors. There are many potential causes for a dam failure. These factors include design error, local geological instability, landslides that cause rapid changes in reservoir pressure, poor maintenance, extreme rainfall, and human or computer error in regulating the water level and pressure in the reservoir.

The Aswan High Dam's Archeological Devastation

Egypt is one of the best-known countries for its ancient history, and the Nubia region along the Nile River in southern Egypt is home to many of these famous sites. Nubia is also home to the largest infrastructure project Egypt ever undertook, the Aswan High Dam, built from 1960 to 1970 (Image 7.25). When finished, the dam provided 50 percent of Egypt's electricity (now accounting for around 15 percent because of a growing population and demand), along with water for irrigation, and brought an end to the annual Nile floods that set the way of life in the Nile valley since the first settlements were built there ten thousand years ago.

The Aswan High Dam (Image 7.26), located near the city of Aswan, is one of the world's more significant hydroelectric power stations. The dam is 111 meters tall (364 feet) and created an enormous reservoir, Lake Nasser, which flooded 5,250 square kilometers (2,030 square miles) of the Nile Valley in Egypt and Sudan. The flooded region covers much of historic Nubia, a region that was central to ancient Egypt and its neighbors.

Although noteworthy for many impacts, it was the destruction of many archeological treasures that captured the world's attention at Aswan. After the dam was started, it was realized that thousands of sites would be flooded, including the famous rock-cut temple of Ramses the Great at Abu Simbel. The United Nations Educational, Scientific and Cultural Organization was called on to save some of the most iconic monuments in Nubia in response the dam's construction.

The most urgent site to save, Abu Simbel, required cutting apart temples cut into the rock cliffs of the Nile and lifting the temples up to a site above the rapidly filling reservoir (Image 7.27). The United Nations Educational, Scientific and Cultural Organization led this effort to deconstruct and rebuild the temples into a fake concrete cliff built directly above the old temple location. Despite modern technologies, the solar calendar inside the temple no longer works on the correct day and the scars of reconstruction are easy to see.

All along Lake Nasser and further afield, other ancient sites have been relocated beyond the manmade floodwaters. Only twenty-two sites were saved, and thousands of sites that are essential to understanding the

Image 7.25 The iconic Abu Simbel temple, carved into sandstone cliffs along the Nile River in Egypt, had to be carved out of the rock and rebuilt when floodwaters forming Lake Nasser threatened to put the temple under water.

Image 7.26 The Aswan High Dam demonstrates that impacts from damming a river can harm history by flooding archeological sites of great importance.

Image 7.27 It took a huge international effort, led by the United Nations Educational, Scientific, and Cultural Organization, to move the famous temples.

Per-Olow Anderson / Wikimedia Commons / Forskning & Framsteg / CC-BY-SA-3.0

early history of civilization are now deep under water and silt deposits that are rapidly filling within the reservoir. Some of the saved temples were moved far away to countries that helped pay to save the most famous sites, including the Temple of Dendur at the Metropolitan Museum of Art in New York (Image 7.28).

continued

continued

Image 7.28 Several temples that would have been inundated by Lake Nasser were given as gifts to countries that helped save Abu Simbel. The Temple of Dendur, now at the Metropolitan Museum of Art in New York City, was given to the United States for its assistance.

Like many other dams, the impacts have gone far beyond a single issue. The Aswan High Dam has had numerous upstream and downstream impacts. Over one hundred thousand people in Egypt and Sudan were displaced from their homes in the Nile Valley. In addition, the dam ended beneficial annual flooding in Egypt that fertilized farm fields and removed excess salts from the topsoil. Groundwater intrusion also is impacting locations hundreds of miles away, including the Great Sphinx at the pyramids in Giza. The dam has led to outbreaks in Nubia of schistosomiasis, a disease caused by snails in stagnant water. Even the Nile delta is paying the price for the dam; it is receding because of a loss of sediments, causing the loss of some of Egypt's best farmland.

Although promised as a savior for Egypt that would provide massive amounts of electricity and irrigation water, the Aswan High Dam has instead destroyed some of the important sites of Egypt's primary tourism sector, while at the same time having many unintended consequences, both upstream and downstream of the dam.

Any of these factors can lead to undue stresses on a dam, which can catastrophically fail, releasing massive quantities of water that rush downstream in a flood (Image 7.29).

The largest dam failure in history occurred in 1975 at the hydroelectric Banqiao Dam in China, when the reservoir overtopped the dam during heavy rains. The dam failure resulted in over 145,000 deaths from the floods and subsequent crop failures, along with over eleven million people losing their homes. Other dam failures throughout history have also captured attention and led to changes in how dams are built and regulated, including the famous Johnstown flood of 1889 in Pennsylvania, which killed thousands (Image 7.30). Although it was not a hydroelectric project, the failure impacted future hydroelectric dam construction in the United States and led to federal regulation of dam construction and liability laws.

Another problem comes at the time of dam construction because the creation of a reservoir can lead to massive population displacement. Several dam projects in China and India have displaced millions of people, necessitating the removal of entire cities. In both

Image 7.29 A partial failure of the dam at Sayano-Shushenskaya Dam in Siberia led to seventy deaths when high-pressure water broke through poorly maintained penstocks and flooded the generator hall in the power plant. Courtesy of RIA Novosti / Science Source.

Ria Novosti / Science Source

Image 7.30 The most famous dam failure in the United States resulted in the 1889 Johnstown flood in Pennsylvania. Although it was not a hydroelectric dam, such accidents could also occur if a major hydroelectric dam failed. Courtesy of New York Public Library / Science Source.

New York Public Library / Science Source

India and Brazil, proposed dams have led to massive protests by people who live in the area where the proposed reservoir would be built. The Belo Monte Dam in Brazil became especially controversial in this regard because native indigenous tribes stood to lose much of their ancestral land to the dam. Over the past century, somewhere between forty and eighty million people have been displaced by dam construction.

Other sites of historical and natural value have disappeared under the water as a result of dam construction. Archeological sites of great value disappeared in Egypt, China, and the United States under major reservoirs. Beautiful natural wonders, including canyons and other landscapes, have been submerged in reservoirs worldwide as well. Although not economically quantifiable in some cases, these losses are great to both nature and human history.

VIGNETTE

Glen Canyon Dam and the Future of American Hydropower

Two dams and their accompanying reservoirs along the Colorado River best represent some of the challenges that will impact hydropower in an era of rapidly changing climate. Glen Canyon Dam and Lake Powell lie just upriver of the Grand Canyon of the Colorado River (Image 7.31), while the more famous Hoover Dam and its accompanying Lake Mead lie just downstream. These two enormous reservoirs, which are the largest by capacity in the United States, lie in the midst of the southwestern deserts of the United States. Thus, the water they capture comes from far away in the higher Rocky Mountains, but all in a part of the United States that is far drier than it was when water amounts were allocated to the southwest states and Mexico in the 1922 Colorado River Compact and is getting drier still.

The Hoover Dam was built during the Great Depression of the 1930s and Glen Canyon Dam was built in the 1950s and 1960s. Together, they provide important sources of electricity and water for use in agriculture and

Image 7.31 The Glen Canyon Dam and its reservoir, Lake Powell, lie just upstream of the Grand Canyon.
Lorcel / Shutterstock

for cities in Arizona, Utah, Colorado, California, and Nevada. They also exemplify the environmental impacts of hydropower in several ways.

Glen Canyon Dam is located just upstream of Grand Canyon National Park (Image 7.32), one of the iconic environmental wonders of the American West. The dam has led to a loss of sediments in the river as it flows through the canyon, which impacted aquatic plants that rely on sandbars and also fish species, and it even slowed the rate at which the river carves the canyon. Fish have been impacted not only by the lack of sediment, but also by the significantly colder water that now flows from deep in Lake Powell through the dam compared with the much warmer water that was heated by the desert sun and was shallow in the river.

Even more significant are the challenges from the amount of water flowing through the Colorado River system, which is lower than envisioned when the compact was created and the dams were built. The years that the compact is based on were the wettest in the past one hundred years in the Colorado River basin. Recent decades, especially the past two, have been far drier, leading to a much lower amount of water flowing into the reservoirs. Both now stand at just a little over half what their capacities call for.

This change is important because a few more dry years could result in the need to choose between the two dams. Otherwise, the reservoir levels could fall to dead pool, at which point the water level would lie below the intakes on the dams. Not only would the electricity supply from the dams be lost, but also downstream flow of the Colorado River would almost dry up (which already occurs a few miles from its mouth on the Gulf of California because of the enormous withdrawals of water in California's southern counties). The loss of hundreds of miles of the river would destroy major agricultural areas and damage the environment in massive ways.

Thus, one of the two dams may have to go. Why Glen Canyon and not Hoover? The answer lies primarily in the location of the dams and their history. On the one hand, the Hoover Dam provides power and water to Las Vegas, the biggest city in the region. In addition, it is a National Historic Landmark and one of the most famous dams in the world. Glen Canyon, on the other hand, lies in a much more remote location. In addition, Lake Powell flooded Glen Canyon itself, which, if the dam was removed, would emerge again from the waters,

Image 7.32 The Grand Canyon of the Colorado River is one of the most iconic national parks in the United States.

Sumikophoto / Shutterstock

continued

continued

revealing a beautiful sandstone canyon that could serve as a recreational area itself, as the reservoir does now (Image 7.33).

Removal of Glen Canyon Dam would also improve the ecosystem of the Grand Canyon by returning higher flow and warmer water to the canyon. The debates will be fierce if this point of no return is reached, but many experts expect that the dam will have to go in the coming few decades.

Image 7.33 Reduced precipitation in the Colorado River basin has led to drops in reservoirs, including at Lake Powell. The "bathtub ring" shows areas that are now above the water level that used to be in the reservoir.
Pmphoto / Shutterstock

Tidal and Wave Power

New technologies have extended the applicability of hydropower beyond river systems to take advantage of tides and currents in open water (Image 7.34). These developments remain mostly experimental, but could be a growing area for future renewable energy development.

Tides, the twice-daily change in coastal water level caused by the gravitational pull of the moon, are one of the few potential energy resources not driven by solar energy. Given the ubiquity and predictability of tides, several types of technologies have been developed in an effort to harvest tidal power. Three major types of tidal power are being developed:

- **Tidal barrage**—a dam is built at an inlet so that water flows over it at high tide and back out through penstocks and turbines as the tide goes back out.
- **Tidal fence**—this system forces tidal flow through narrow slits, which then turn blades in an area that experiences large differences between high and low tide.
- **Tidal turbine**—this technology works like a wind turbine, with tidal flow turning an underwater propeller (Image 7.35).

Image 7.34 The Bay of Fundy has the largest tidal variation in the world.
Alberto Loyo / Shutterstock

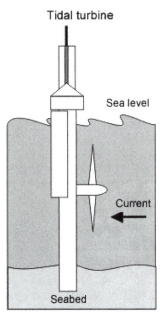

Image 7.35 Tidal turbines work similar to wind turbines, but are placed in areas of tidal flow near shore and underwater.
Courtesy of the U.S. Department of Energy

All these technologies are hampered by the twice-daily nature of tides. This causes power to be generated intermittently, with barrages only producing power as the tide subsides from high to low, while the other technologies produce most of their power during the most pronounced transition periods between the tides (Image 7.36).

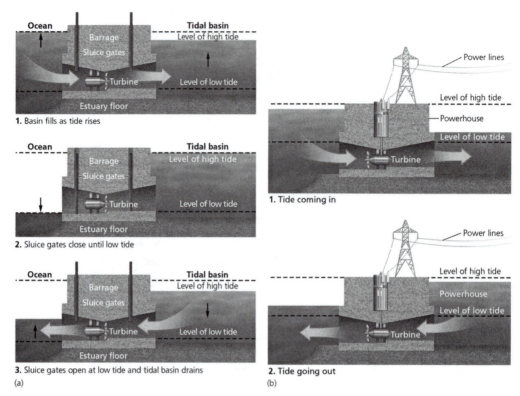

Image 7.36 Tidal power stations function by either (a) allowing a basin to fill at high tide, where it can be stored until low tide and then released, or (b) letting the flow of water between high and low tide power a turbine as the tide changes.

Another potential power source would take advantage of the energy in waves. Unlike tides, waves are wind driven and therefore are originally derived from the sun. Waves vary in intensity and frequency in different coastal areas, making the choice of siting for wave power critical. Wave power works similar to a tidal barrage by channeling water from large waves into a reservoir on the coast, where the water can then flow back out through penstocks and turbines. Because of the scale needed, this method works only in places with large waves. There have been proposals to build wave turbines (similar to the tidal turbine pictured in Image 7.35) in areas with consistent currents and waves, which would in essence be similar to a wind farm. To date, no such technology has been implemented on anything but an experimental scale. The dropping costs of offshore wind farms, combined with their reliability and already developed technology, make them a more attractive alternative in areas where constant wind would generate sufficient wave power. Despite this, research in developing more unconventional hydropower technologies continues.

REVIEW QUESTIONS

1. Other than for the production of electricity, how can hydropower be used?
2. What factors lead to hydropower having variable ability to meet energy needs?
3. Why are only some river systems well positioned to provide large-scale hydropower?
4. How can small-scale hydropower be used in areas not suited for large systems?
5. Why do most dams have a short life span of one hundred years or less?
6. Other than natural impacts, what are social and political negative impacts of hydropower?
7. Why is the U.S. government a big player in hydro projects?
8. What are the basic differences of the three tidal power schemes?
9. Where does tidal power gain its energy, since it is not solar driven?

Image 8.1

CHAPTER 8

Wind Energy

Wind power has been appropriated for millennia as an energy resource. Until the late nineteenth century, wind drove much of the world's trade and naval power through sailing ships. Sailing technology had been in use for thousands of years, with the Egyptians developing sails by at least 3500 BCE (Image 8.2). By the nineteenth century, some of these vessels could harness up to ten thousand horsepower from their sails! Until coal-burning steam engines began to drive both shipping and the growing railroad networks across the world, **wind** was one of the most crucial energy resources to be harvested by civilization.

Windmills have also been in use for many centuries. Early **water pumps** utilizing a vertical axis design may have been in use in China, Tibet, and Afghanistan by the sixth century CE. The first horizontal axis designs ground cereal crops into flour in Persia (modern-day Iran) and made their way to Europe during the Crusades (approximately 1100 to 1300 CE). Later, windmills were refined in the Netherlands and northern Germany to pump water, long before being used to generate electricity. The Dutch designs utilized the propeller shape that has become standard in the design of most modern **wind turbines** (Image 8.3). By the start of the twentieth century, millions of windmills pumped water in the United States alone, mostly on farms and ranches in isolated areas (Image 8.4). Early electrical generation wind turbines served the western United States in isolated rural areas to power radios, light bulbs, and other basic electronics. Similarly, Denmark deployed electrical wind turbines in the 1920s on isolated islands that otherwise had no electricity.

Image 8.2 Wind power once was important for transportation, such as its use on sailing ships.
Currier & Ives., Library of Congress

Image 8.3 Windmills were used in the Netherlands to pump water and reclaim areas of land from the sea.
Natali Glado / Shutterstock

Despite its long history as a key resource, there has only been interest in developing larger scale **wind power generation** since the 1970s. There have been major strides in developing wind energy for large-scale electrical generation in the past two decades, in terms of both improved technology and lower cost. In fact, in a good windy area, new wind farms provide the lowest-cost electricity of any other generation technology.

Image 8.4 Small windmills were used in the American West to pump water from underground for use on farms and ranches.

Greg Fugate / Shutterstock

Image 8.5 Windmills have been in use for small-scale industry for thousands of years, such as these windmills that powered small cottage industries in Mykonos, Greece.

This has led wind to become an increasingly important part of the global energy basket as a key renewable energy resource. Today, wind farms are often seen as a symbol of progress and modernity, rather than the latest iteration of an energy resource that has been in use since early human history.

The Wind

Winds derive their power from a combination of energy from the sun and the rotation of the earth. Because of the uneven heating of the earth's surface, air currents circulate throughout the atmosphere in the redistribution of that solar heating. Wind patterns are both local and global, with local winds driven mainly by surface terrain (winds blow with less restriction over plains than they do over mountains or forests, for example) while global wind patterns are generated by the temperature differences between polar and equatorial areas, as well as the rotation of the earth around its axis.

Different surface types (land vs. water) account for localized uneven heating, leading to a concentration of strong winds in many coastal environments. During the day, land surfaces heat faster than the water because of the lower amount of energy needed (specific heat capacity) to heat the land. This leads to the temperature of the air over land rising faster, to be replaced by cooler wind coming from the sea. After sunset, the land cools much faster, leading to a reversal of the pattern, with the wind blowing from land to sea at night (Image 8.6). Local wind patterns are further

A Sea breeze

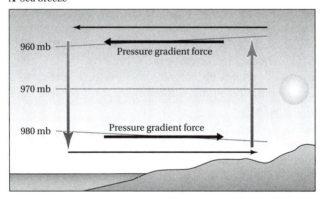

B Land breeze

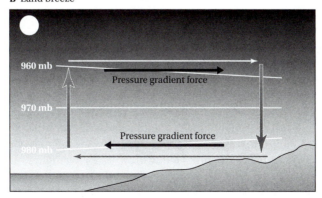

Image 8.6 During the day, sea breezes blow from over the water toward land and reverse at night. Wind is more constant in coastal areas than in many areas inland.

influenced by other factors including terrain, vegetative cover, and local weather patterns. In certain U.S. states, such as Wyoming and North Dakota, there is little ground cover and extensive wind belts experience consistently high wind speeds during the whole year.

In other places, wind patterns can vary seasonally. In Southern California, winds blow toward the Mojave Desert from the Pacific Ocean more strongly during the summer, when desert air is hottest and rises quickly. This leads to more wind energy potential at the same time of year when electricity demand for air conditioning is highest. Similarly, the winds blow strongly across the Great Plains of the United States during the summer months, as air flows from the Gulf of Mexico and Arctic collide across the region, producing storms that can include tornadoes. Offshore, wind patterns are even stronger and more consistent, although offshore turbines are more difficult and expensive to develop.

Wind's variability over short periods can prove challenging. A few unique points on the earth's surface have high sustained winds at all times. However, most locations see tremendous variation in wind depending on the season, time of day, and vagaries of weather. Unlike solar power generation, wind does have the advantage that it blows at night, so the resource is not limited by time of day. While not limited by day/night cycling, wind power is limited by the larger patterns determining when wind blows, including daily variations in weather pattern. Despite those variations, wind is very consistent over broader time periods, so daily variability is compensated on a larger scale by the predictability of wind over weeks, months, and years.

Before a site is selected for wind generation, several factors must be considered. The U.S. Department of Energy recommends the following key considerations for locating a wind power facility: good wind resource, adequate transmission lines, reasonable road access (for onshore sites only), ability to obtain permitting, a receptive community, and few environmental concerns. Many local factors influence wind sites as well, with flat or remote areas having stronger wind than places with many such obstacles as trees, buildings, or hills. Because of this complexity, each site must be selected based on its own conditions and the needs of the power grid.

Wind Turbines

Modern wind energy generation is conducted with two types of turbines—horizontal axis and vertical axis. The horizontal-axis turbine is both the most well known and the most commonly used type of turbine and has come to effectively dominate wind generation at commercial scales (Image 8.7). It resembles a large airplane propeller and works on a similar principle, with the blades facing into the wind. On an airplane, there is a motor driving the propeller that blows air to fly the plane. On a wind generator, the wind pushes the propeller, which drives a generator at the center to harness the kinetic energy of the wind. A standard horizontal-axis turbine usually has three blades and can stand well over 100 meters tall, with blades over 80 meters long on the largest models. Horizontal turbines have the advantage of great reliability

Image 8.7 Most modern windmills use a horizontal axis.
Dario Racane / Shutterstock

and relatively low cost. Because they have been by far the most common type of unit produced, they have seen the greatest benefits of engineering improvements and economies of scale in manufacturing, making them the most cost-efficient turbine type available.

Vertical-axis turbines, also known as **Darrieus turbines**, look more like an egg-beater with two vertical blades (Image 8.8). Almost all currently sold units are small and used for local generation, but the towers of some larger models span up to 30 meters high and 15 meters wide. Only a small percentage of wind turbines currently in use are of the vertical-axis type. Despite the predominance of horizontal turbines, vertical units have a number of advantages in certain circumstances. Vertical turbines are more efficient at lower wind speeds, allowing them to be efficient at small size while closer to the ground, and they do not need to face the wind directly, eliminating the need to rotate the unit. The complexity of vertical turbine units has meant that they are less reliable and relatively rare compared to horizontal-axis wind turbines.

Electrical power generation by wind power is relatively simple. The blades turn a shaft that is connected to a turbine (Image 8.9). Electrical generation is identical to other mechanical and fuel energy sources, with the rotation of electromagnets within a generator converting mechanical energy to electricity. On horizontal-axis turbines, the unit is located behind the blades in a city bus–sized nacelle, high above the ground, while on vertical-axis turbines, it is located beneath the blades and close to the ground. A low-speed shaft connects directly to the blades and rotor and usually rotates about thirty to sixty times per minute. This leads to a gearbox that turns a high-speed shaft, which then turns the electromagnets inside the generator. Additionally, the turbine measures wind speed and direction using an **anemometer**, which helps the turbine controller maximize electricity generation. Most large turbines start to work at wind speeds above 13 kilometers per hour (kph) and stop working to protect the mechanical

Image 8.8 Some windmills use a vertical axis instead.

Lenise Calleja / Shutterstock

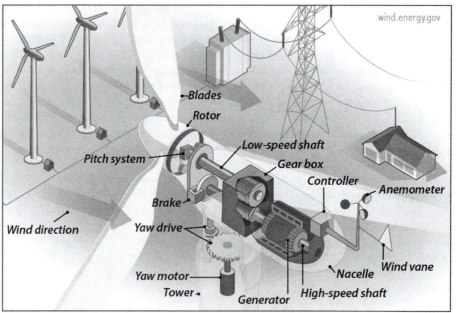

Image 8.9 Wind turbines use the wind to turn blades, which then turn a crankshaft, which in turn powers an electrical generator.

U.S. Department of Energy

components at speeds above about 90 kph. In cases of emergency or high wind, the shaft is connected to a brake to stop the blades. On horizontal-axis turbines, there is also a yaw drive and yaw motor, which turn the turbine to face into the wind. On small residential turbines, there is a tail that rotates the unit in the same manner as a weathervane, rather than the yaw controller. Some turbines also have drives that change the pitch of the blades to turn the blades more or less into the wind, further maximizing power output or limiting stress on the brake during high-wind events.

Wind turbine technology has developed quickly in recent years, with major advances in efficiency and a contemporaneous drop in production costs. Some modern turbines convert wind energy into rotational energy at a 50 percent efficiency, and it is theorized that no wind turbine could attain better than 59 percent efficiency because of limitations in fluid mechanics. Further loss in the conversion of mechanical energy to electric energy leads the most efficient wind turbines to transform no more than slightly over 40 percent of the kinetic energy from the wind into electrical energy at the turbine's location. Current technological development increases efficiency by 2–3 percent annually in new designs. Most of this improvement stems from increasing hub heights (wind is more consistently strong higher above the ground) and larger blade lengths.

Wind turbines come in a large variety of sizes to meet different needs. Small-scale turbines are generally classified as those generating 10 kW or less of power (Image 8.10). These are used primarily for individual homes, farms, and remote applications such as telecommunications towers or water pumps. Small-scale turbines are relatively cheap solutions for off-grid use, with models running about $2,500 to $8,000 per kilowatt, or costing about $0.06 to $0.26 per kilowatt-hour. At such prices, they are not cost-effective for larger scale usage. Most small-scale turbines require an average wind speed of 14.5 kph to be effective.

Intermediate-scale turbines generate between 10 and 250 kW of power (Image 8.11). These turbines can be used in schemes for small village power generation and in some

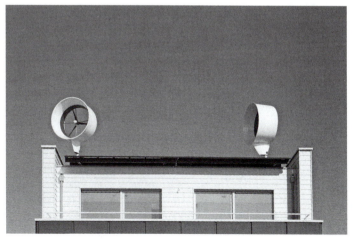

Image 8.10
Worldpics / Shutterstock

Image 8.11
CoolKengzz / Shutterstock

cases are tied into broader distributed power systems that also include electricity generated from other sources.

Large-scale turbines produce over 250 kW of power, with the largest current prototype generating 10 MW of power (Image 8.12). Large-scale turbines are almost always part of larger distributed power networks operated by utilities. Because of their size and cost, they are not usually part of power schemes below such levels of scale and distribution. Large turbines cost closer to $1,700 per kilowatt, or a minimum of over $250,000 to erect. However, because of their economies of scale, the cost per kilowatt-hour ranges from $0.02 to $0.06. Most large turbines require an average wind speed of at least 6 meters per second to be effective.

Although turbines can be erected alone, most large-scale wind turbines are arrayed into **wind farms**, either on land or offshore (Image 8.13). Wind farms are best located in areas of consistently moderate to strong winds to ensure consistent generation, leaving many other areas as poor candidates for wind farm development. Site selection is key to ensure a successful project. Sites without extensive tree cover, such as on plains or atop hills, are ideal, as are sites located at gaps in a mountain chain where winds funnel through. Additionally, sites near the shoreline can take advantage of the shifting winds between land and sea. Concentrating generation is important to save on construction and grid-integration costs because most wind sites are located far from population centers that use the amounts of electricity that a big farm will generate.

Image 8.12
Stillfx / Shutterstock

Image 8.13 Large utility-scale windfarms were first built in the United States in the 1970s, such as this windfarm outside Palm Springs, California.

Often, new high-capacity transmission lines must be constructed to connect a wind farm to the power grid.

Offshore sites add some specific advantages and drawbacks (Image 8.14). One advantage is that they can be easily located close to major population centers, since most human populations concentrate near the coast. Offshore wind, in general, faces

Image 8.14 Offshore windfarms take advantage of the more consistent wind patterns over large bodies of water, such as this windfarm off the coast of Denmark.

Tony Moran / Shutterstock

less of a "**not-in-my-backyard**" (NIMBY) problem than onshore sites because local communities do not generally live close enough to the power stations to oppose their construction. Additionally, most offshore wind farms generate for more hours of the year because the wind is generally much more reliable offshore than over land. Off-shore wind farms do face additional challenges, however, particularly in cost. In water 9 meters deep, the costs are already 30 to 50 percent higher than onshore because of the need for more extensive foundations to resist waves and violent sea storms, which are usually stronger than similar storms on land. Additionally, the costs to connect to the grid are exacerbated offshore, especially if the systems are more than 10 kilometers from the coastline. Finally, offshore turbines must be of higher reliability than their on-shore cousins because of the higher cost and difficulty of maintaining them.

The rapid development of wind energy has led to the construction of some large wind farms across the globe. The world's largest is the Gansu wind farm, located in western China. It is home to over 3,500 wind turbines and can generate over 5,000 MW of electricity. The largest offshore wind farm is the London Array located in the Thames River estuary off the east coast of England. It has 341 turbines and a maximum capacity of 630 MW, with plans for 1,000 MW capacity once its second phase is completed.

As noted before, wind variability from day to day, or even hour to hour, can cause a challenge for wind power integration into a larger grid. This makes site location for wind farms essential, to ensure that they stand in places where winds are consistent. Even so, an electrical power system that uses wind power must have backup energy sources in place for windless times. Much of the variability in wind can be eliminated by having a grid sourced by many wind farms in different places because it is highly un-likely that all will be idle at the same time. At present, fossil fuel generation is the most common backup solution, but a more diverse renewable grid or the inclusion of battery or other storage will likely become a more commonplace measure in the near future.

Worldwide Wind Power Generation

Wind power has expanded dramatically over the past twenty years. Concern over climate change and falling costs for wind electricity have led to rapid increases in wind power generation. Cost has become the main driver of wind most recently because the price of a kilowatt-hour of wind has reached or even gone below that of conventional energy sources in many places. Growth in China is being driven by Chinese desire to diversify their energy basket and to reduce air pollution from electric generation. Wind is the fastest-growing renewable source in the developed countries of the world and has been catching on in some developing regions as well. India has massively increased its investment in wind generation, and wind power projects have also gained traction in such diverse countries as Costa Rica and Kenya. In regions such as sub-Saharan Africa, wind power is rare, but may become more common as electricity demand increases and wind energy costs continue to decrease.

The top five wind producers in 2014 globally were China, the United States, Germany, Spain, and India (Table 8.1). As a percentage of total electricity generation, Denmark is the leading wind power producer in the world, with over 39 percent of its total electricity coming from wind, mostly from large offshore wind farms. The Danish company Vestas is also the world's leader in the production of wind turbines. In Denmark, some politicians have even called for moving to 100 percent wind-powered electricity. However, because of variability in wind, this could be problematic for the sustainability of their power grid without other supplemental systems or major advances in energy storage systems.

Table 8.1 Top Ten World Wind Generators (megawatts installed capacity)

China	168,690
United States	82,184
Germany	50,018
India	28,700
Spain	23,074
United Kingdom	14,543
France	12,066
Canada	11,900
Brazil	10,740
Italy	9,257
World total	486,749

Source: Global Wind Energy Council, 2017, Global Wind Statistics 2016, http://www.gwec.net/wp-content/uploads/vip/GWEC_PRstats2016_EN_WEB.pdf.

Wind power's success has partly been based on the ease of installation for communities and smaller utility companies. In southern Europe, especially Spain, Portugal, Italy, and Greece, wind power has boomed despite the economic crisis in the region. If anything, the crisis has helped spur wind development. Electricity rates were previously especially high in these countries, and with the lowering cost of wind power, it has proven attractive. A good case study is the small Italian town of Tocco, which invested in wind turbines to offset high local electricity prices. The town has been so successful that it now exports electricity to the region's power grid and has been able to reduce taxes based on this income. Small wind schemes across these countries have revolutionized the manner in which electricity is produced by giving small producers a role in the national electrical grid. European **feed-in tariffs** have

VIGNETTE
Birds and Wind Power

During the first boom of wind turbine construction in the 1970s and 1980s, studies began to show a major uptick in the number of birds dying from striking turbine blades at wind farm sites. The studies demonstrated an especially large increase in mortality among large bird species, such as eagles and vultures. However, more recent studies show that older models of wind turbines were more problematic and that earlier wind farms were not given appropriate site evaluations as to their location along known migratory bird routes.

Over time, scientists have worked to improve understanding of how wind power sites affect bird populations and, more recently, bat populations. Obtaining exact data is difficult because many factors beyond strikes on wind turbines play into avian mortality figures, making it hard to quantify numbers accurately. Furthermore, it is even tougher to compare the impact of wind farms to that of other manmade hazards that impact birds. By 2005, the U.S. Fish and Wildlife Service estimated that wind farms were killing more than 40,000 birds annually in the United States. This compares to somewhere between 4 and 50 million birds killed in strikes on communication and radio towers across the country and somewhere between hundreds of thousands and 175 million fatalities from collisions with power lines, demonstrating how difficult it is to gather accurate data on this problem. Wind farms did not appear to be a leading cause of mortality, but some California studies showed a much larger percentage of the fatalities affecting large birds, which play a disproportionate role in local ecosystems.

One wind facility, the Altamont Pass Wind Resource Area, located east of San Francisco, California, came under special scrutiny (Image 8.15). By the early 1990s, the 6,500 turbines at this facility were killing several hundred raptors each year, including golden eagles, red-tailed hawks, and American kestrels, all of which are major predatory species that control populations of rodents and reptiles (Image 8.16). These concerns led to the replacement of many fast-moving smaller turbines with larger scale turbines that are less harmful to birds. This decreased total fatalities, but it was still estimated that forty to sixty golden eagles were killed annually, a disturbingly high number for a primary predatory species that is protected under the Endangered Species Act. The initial problems with Altamont have greatly helped other projects avoid many of the same problems, however.

More recently, major concerns have been raised over bat strikes on turbines in some areas. It is believed that the pressure differentials caused by the blades confuse bats' echolocation (use of sound to navigate), making

Image 8.15 The Altamont Pass windfarm in California revealed that windmills can kill large numbers of migratory birds.

Martin Vonka / Shutterstock

Image 8.16 Golden eagles and other rare species of birds must be considered in the locating of windfarms.

them more likely to strike the blades. Additionally, the lights on large turbines that are needed for airplane safety may attract insects. Because bats are a major controller of insect populations and major pollinator of food crops, this is of special concern.

New forms of monitoring are being developed to help better understand the problems associated with bird and bat strikes, so that wind farms can be better designed and equipped to minimize losses. Although wind turbines make up a small percentage of total bird fatalities, the high incidence of raptor deaths and the problems with bat strikes merit continued study so this green technology does not devastate a key part of the ecosystem and food chain. The lessons from early wind farms like Altamont have helped wind energy developers greatly redesign later projects to improve bird and bat safety; the turbines themselves are designed with these goals in mind, and a more comprehensive approach to site location can also greatly improve bird safety. Several major environmental groups, such as the Sierra Club and the World Wildlife Federation, now support carefully designed wind farms.

further spurred this development, both in the south and in other regions of the European Union. The tariff guarantees a minimum price for the power when sold to the power grid that reflects the cost of its generation, regardless of the prevailing electricity price. This policy has been instrumental to European wind power development; however, it has also caused problems for some countries, notably Spain, during the economic crisis.

Wind Power Generation in the United States

The geography of the United States provides thousands of prime locations for the generation of wind power (Image 8.17). The U.S. potential for onshore electricity generation, excluding areas unlikely to be developed, is 38 million GWh, or about ten times the total U.S. electricity generation from all sources. Onshore, the Great Plains region

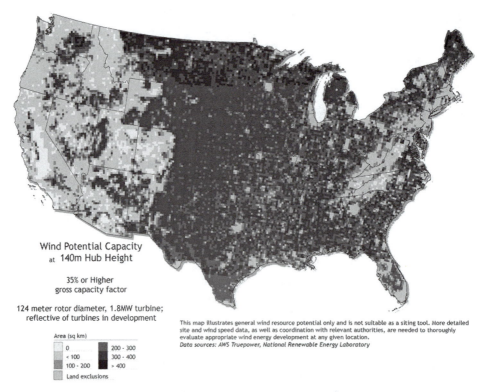

Wind Potential Capacity
at 140m Hub Height

35% or Higher
gross capacity factor

124 meter rotor diameter, 1.8MW turbine;
reflective of turbines in development

Area (sq km)

	0		200 - 300
	< 100		300 - 400
	100 - 200		> 400
	Land exclusions		

This map illustrates general wind resource potential only and is not suitable as a siting tool. More detailed site and wind speed data, as well as coordination with relevant authorities, are needed to thoroughly evaluate appropriate wind energy development at any given location.
Data sources: AWS Truepower, National Renewable Energy Laboratory

Image 8.17 Different parts of the United States have varying average wind speeds. Onshore, many of the best areas for wind farms are located in the Great Plains and Rocky Mountain states.

Courtesy of the U.S. Energy Information Administration, adapted from the National Renewable Energy Laboratory and AWS Truepower

of the United States has the greatest average wind speeds, with large swaths of states from North Dakota to Texas containing areas with the most consistent wind speeds. Conversely, the South is the region with the poorest wind resources. Offshore, there is significant potential for wind development, with most of the East Coast and Northern California having the greatest potential. Because of the broader and shallower continental shelf, East Coast projects will be easier to develop, and they also happen to be located much closer to major population centers, which provides the benefit of producing power where transmission loss can be minimized.

According to the U.S. Department of Energy, wind power is the fastest growing source of electrical generation in the United States and has been for several years. It now makes up almost 4.5 percent of total electrical generation in the United States and 26.9 percent of renewable energy generation. The rise of wind developed quickly, with wind power making up a smaller portion of renewable electricity generation than biomass, waste, or solar in the year 2000 and constituting a larger share than all of them combined by 2007. In the year 2000, only 2,539 MW of wind power generation was installed in the United States, with over half of all capacity in California. Since the 1970s, California had been the largest producing state until it was surpassed by Texas in 2005.

By the end of 2014, the situation changed dramatically, with total U.S. wind capacity at 65,879 MW, an increase of over twenty-five times the amount just fourteen years prior. This shift has taken place across the country, and Texas now leads the United States in total installed wind capacity, with 12,212 MW and a total generation of 35.8 GWh in 2013. The fastest growth in new commissions for wind generation occurred in 2012, following a slowdown during the global economic crisis. As with all forms of energy, tax incentives and subsidies have promoted wind generation. A loss in these subsidies would likely slow further deployment of wind turbines, but not halt them, because the startup costs have dropped by over 40 percent just from 2009 to 2012 as the number of wind turbines produced has grown (Image 8.18). Improvements in the efficiency of turbines, allowing them to produce power at lower wind speeds, will also abet continued wind development once subsidies are scaled back or eliminated. Utilities have also used **voluntary pricing schemes** to support wind development, where some customers pay extra to the utility to purchase "**green power,**" whereby proceeds are diverted to fund additional renewable energy projects. Over 350 utilities have launched such programs, but participation rates remain low.

Once offshore wind farms begin to operate in the United States, there will be a further change in the overall wind power makeup, with larger portions of wind power coming from the Northeast and mid-Atlantic, where most of the proposed offshore wind farms are likely to be constructed in coming years.

Environmental and Political Impacts

Wind power provides many tangible benefits for the environment and reduced costs, which have led to growing interest in its development. Wind is a proven renewable energy resource that is coming down in price as it develops. Wind is a zero-emission

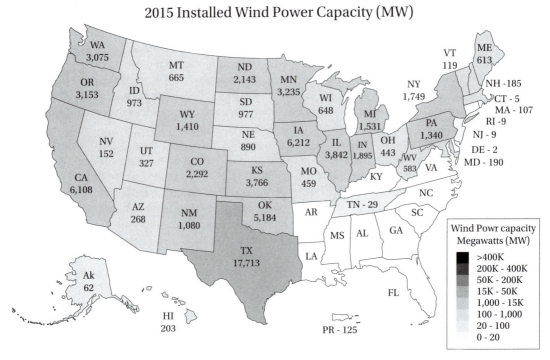

2015 Installed Wind Power Capacity (MW)

Image 8.18 Texas leads the United States with the most installed wind capacity, with more than three times that of California, the state with the second highest capacity.

Courtesy of the Office of Energy Efficiency & Renewable Energy

resource, and it does not contribute to climate change. These advantages have led a broad variety of groups to support the expansion of wind power and have been a driving force in the aggressive development of wind power resources throughout the world. Wind variability must be managed, but a diverse power grid can ameliorate many of the problems with variability of most renewable energy sources.

Another often-cited problem is bird strikes with wind turbines. However, according to the U.S. Fish and Wildlife Service, only about 33,000 birds die in the United States each year from such strikes, compared to somewhere between 97 and 970 million bird deaths from striking glass windows on buildings. Although a casualty figure of 33,000 should not be taken lightly, compared to bird collisions with buildings, radio towers, cars, and high-tension power lines, strikes on wind turbines are rare and account for far under 1 percent of such deaths.

Some of the most organized political opposition to wind farms has come from people who live near them and complain of noise. In most cases, the sounds detectable from large wind turbines come primarily from the blades. Rare are complaints about the sounds generated in the nacelles and gear boxes atop a wind turbine, whose noises are muffled by the design of the structure. There have been a growing number of lawsuits brought to close down wind farms in many U.S. states, including Illinois, Texas, Massachusetts, and Pennsylvania. Similar complaints have been lodged in Canada, the United Kingdom, and France. In some of these suits, people have claimed that the wind turbines cause physiological problems such as irregular heartbeat, blurred vision, nausea, and sleeping disorders. These complaints have been limited to a small number of wind farm sites, and the majority have not experienced such objections. There has also been a deficit of scientific backing to the more extreme of these claims. While some large-scale wind turbines create audible sounds from the movement of the blades, these noises are usually hard to detect unless a person is very close to a turbine. In addition, there is likely to be so much wind noise that any sounds the turbines make will be hard to hear. Most lawsuits have failed to gain traction, and U.S. Department of Energy studies show that property values have been unaffected. In some cases, wind farms have been ordered to slow the blades at night to reduce noise, which also leads to reduced electrical generation—a difficult payoff for those investing in wind power generation.

For others, the principal opposition to wind power stems from its appearance. Wind turbines are large and therefore can be seen over long distances (Image 8.19). This is enhanced by the need to place wind turbines in open areas and often on higher ground, thus magnifying their presence. This is an entirely subjective argument because many people view the turbines as ugly, while many others view them as symbols of progress and think they are attractive. Analysis of property values in the United States have so far failed to show any negative effects of wind turbines on property values.

Image 8.19 Wind turbines located near housing can cause not-in-my-backyard issues.
Industry and Travel / Shutterstock

Offshore Wind Debates

In Europe, offshore wind has become a major component of the renewable energy portfolio (Image 8.20). In contrast, the United States has no offshore wind farms, despite a large potential for them.

The great debate in the United States has focused around one major proposal, the Cape Wind project in Massachusetts. Cape Wind is a proposal to build 120 offshore turbines in Nantucket Sound, between the Massachusetts coast and the popular destination islands of Martha's Vineyard and Nantucket (Image 8.21). The project would be at least 8 kilometers from the closest land and 22 kilometers from Nantucket. The project is set in a 62 square kilometer area of shallow water and would produce up to 468 MW of electricity.

Image 8.20 Offshore windfarms have caused great debate in the United States.
Steve Bridge / Shutterstock

Image 8.21 The Cape Wind project would have installed numerous windmills in Nantucket Sound, which is part of a popular area for tourism and is sacred to local Indian tribes.
Marcia Crayton / Shutterstock

Cape Wind was first proposed in 2001 and had to fight for over twelve years through a variety of hurdles before starting construction in late 2013, though the project has once again halted due to legal and regulatory appeals. These hurdles came in the form of complaints both about the project's appearance and about birds, shipping lanes, and maritime regulations. The one thing both sides could agree on is that Nantucket Sound has a consistent wind resource.

The Cape Wind fight focused more on appearance than any other issue and led to a conflict that pitted neighbor against neighbor in discussing the wind farm. The complex would be just visible from the mainland and both heavily touristed islands. For some, the complex of wind turbines looks like progress, while for many others it is an eyesore. The envisioned project would have forty-story-tall turbines that would be clearly seen from the shoreline. Opponents note that these turbines are taller than the Statue of Liberty and ten times taller than any structures currently near the shore of Nantucket Sound.

Another angle that came into play involved indigenous rights, with the Wampanoag tribe opposing the wind farm because of its location in their sacred Horseshoe Shoal area, where they hold sunrise ceremonies. This tribal complaint led to separate lawsuits challenging the installation.

The debate also addressed environmental concerns. Those in opposition pointed to potential bird strikes and damage to the seabed that could affect fisheries and whale migration routes. On the other side of the environmental equation, Cape Wind notes that the project would provide 75 percent of the electrical needs for the area in a clean and **carbon-neutral** fashion. A federal environmental impact study also showed that the impacts to birds and sea life were far less than opponents claimed.

The final federal study compiled over eight hundred pages of information from a vast array of government agencies, concluding that the project should go forward. These studies were challenged in court and upheld with the support of many residents and major environmental organizations like Greenpeace and the Sierra Club. After twelve years of litigation and debate, the project remains stalled as the United States' first offshore wind farm. It is questionable when we will see major offshore wind development along both coasts of the United States and in the Great Lakes.

Although Cape Wind was the first proposed site, in 2013, the first offshore wind turbine in the United States was deployed in Maine (Image 8.22). Unlike Cape Wind and the major European offshore wind farms, this is the first grid-connected offshore floating wind platform in the world to use a concrete composite floating platform, yet another step forward in the development of wind power.

Image 8.22 The first offshore wind turbine in the United States is a floating turbine off the coast of Maine.

Robert Bukaty / Associated Press

A final policy debate that continues to rage regarding wind power (and all renewables more broadly) is over the use of subsidies. Both 2005 U.S. legislation and the 2009 federal stimulus gave extra incentives to install wind power. Some of this incentive comes in the form of direct tax rebates for the installation of wind turbines, while other subsidies guarantee feed-in tariffs, which allow wind power to be sold at a minimum price, regardless of the cost of electricity in an area. The tax subsidies lead to reduced tax revenue, while feed-in tariffs can impact consumers by affecting the price paid for electricity. For many consumers, higher electrical rates are objectionable, even if they are caused by the installation of cleaner energy sources. Like all public policy debates on taxes and spending, the arguments that involve wind power subsidies are politically charged with strong objections on both sides—one arguing that only the free market should determine which sources of power are used, while the other focuses on the cleanliness of the energy and argues that the market must be manipulated to ensure our use of cleaner energy options. Most other forms of energy, including fossil fuels, have their own subsidies as well; consequently, objections to subsidies for renewable energy must be placed in the larger context of the government subsidizing virtually all energy resources through different channels.

Wind power has fast emerged as a major alternative energy resource for electricity generation. Although wind power has a long history, the modern energy era mostly eclipsed wind during the late nineteenth and all of the twentieth century because of the explosion of fossil fuel usage. A combination of newer, more efficient technology, falling costs, and an interest in reducing carbon emissions has changed the course for wind in a dramatic fashion.

Despite this, wind will face many ongoing challenges. The variable nature of wind must be managed, likely in conjunction with other renewable energy development. Efforts must also be made to ameliorate the concerns of potential bird and bat mortality and other environmental impacts of wind power to ensure that it is a truly green power source.

Wind has become a major player in the energy basket of some countries, especially in the European Union. Additional expanding capacity in the United States and China ensures that it will become an even larger player down the road. Wind also offers great hope to more isolated communities that live without electricity as a viable option for small-scale power. In an era of rising standards of living and concern over the global climate, the future for wind power is bright. It will not prove to be the silver bullet to wean humanity off fossil fuels, but will likely play a major role as one of many growing renewable energy sources.

REVIEW QUESTIONS

1. How was wind power an integral resource prior to the modern era?
2. What two factors cause the world's winds? How does this source generate air currents and wind patterns that differ around the globe?

3. What are the key factors for considering a site for wind generation?
4. Identify and explain the key challenges and/or criticisms of wind power generation.
5. Is wind power a total solution to greenhouse gas emissions?
6. Describe the success of southern European countries in implementing wind power generation.
7. What areas of the United States have the greatest wind potential and why?

Image 9.1 Andrei Orlov / Shutterstock

CHAPTER 9

Solar Energy

Almost all forms of energy we consume, whether it be fossil fuels, wind, or hydropower, ultimately derive their energy from the sun. The earth receives massive amounts of solar energy every year—about 23,000 TW-year every year. This is well over ten thousand times the amount of energy humans consume each year, although most of that energy could never be harnessed by mankind. Because of advances in technology, the direct use of solar energy has grown rapidly, and solar power can be used in a wide variety of applications.

As with other renewable resources, solar energy faces many challenges. Like wind, it is a variable resource by time of day, weather, and season. Solar is strongly dependent on geographic location. Generally, the closer you are located to the equator, the more consistent solar power is over the year. Despite these obstacles, technological advances and falling costs have allowed solar to become the fastest growing renewable energy source, and when paired with other energy technologies, it is likely to continue to grow rapidly in the coming years.

Solar energy has been used by humans for a long time. By the seventh century BCE, humans were using magnifying glasses to start fires (Image 9.3), and later the Greeks and Romans used concave mirrors to light torches. The sun's energy has also figured into building design since humans began building. Buildings have been constructed to either capture the sun's heat and light (especially in cool climates) or reflect much of that energy (especially in warm climates) through the orientation of windows, choice of building materials, and other features.

Image 9.2 Solar energy has always been used by humans, such as for simple purposes like producing salt.
Fabio Lamanna / Shutterstock

Image 9.3 Solar energy can be magnified and used to produce heat and even fire.
Sydeen / Shutterstock

Generating electricity directly from sunlight is a relatively recent phenomenon. In 1839, the French physicist Alexandre-Edmond Becquerel first discovered the **photovoltaic effect** (light directly to electricity). The first solar cell was built by the American inventor Charles Fritts in 1883, but was only about 1 percent efficient. The first modern silicon-based photovoltaic (PV) cell was invented in 1954 in the United States by researchers at Bell Telephone Laboratories. In an astonishing example of how rugged solar cells are, those first research cells still work.

Through the 1960s and 1970s, solar cells were mostly used in niche applications, including powering satellites and handheld calculators (Images 9.4 and 9.5). In 1977, world photovoltaic generation hit 500 kW, about enough solar for electricity use of one hundred homes. Finally, in the 1980s, we started to see more widespread use of solar. Major breakthroughs continue to reduce costs and increase efficiency of generating electricity from solar to a point where it has become competitive with other forms of electric generation. However, only about 1 percent of global electric generation today comes directly from solar power.

In this chapter, you will learn about solar used to generate heat and electricity. For electric generation, this will include the generation of electricity through photovoltaics

Image 9.4 Early solar photovoltaic panels were inefficient, but could be used for low-power devices like calculators.

Andy0man / Shutterstock

Image 9.5 Solar energy is more intense beyond the protection of the earth's atmosphere. It is used to power many satellites, including the International Space Station.

Marcel Clemens / Shutterstock

and solar thermal plants. You will understand how solar power fits into the global and U.S. energy system and the impact that solar power generation has on the environment.

Basics of Solar Energy

Today, we use solar energy to produce both heat and electricity. Solar heating is used to directly heat water and the spaces that people use. Solar power is also used to generate electricity in two different ways: **photovoltaics**, which convert sunlight directly into electricity, and **concentrating solar power** (CSP), which harnesses solar heat to produce electricity through thermal generation. These two varieties of solar power generation have a wide range of subtechnologies of greatly varying cost, efficiency, and capacity. Worldwide solar electricity generation has ramped up significantly in recent years, encouraging many innovations in solar energy–capture technology. These developments have only recently made solar PV cost competitive with traditional fossil and nuclear electricity generation.

The many benefits of solar generation must be balanced against its weaknesses, chief of which is the variability of solar energy as a source. New technologies aim to more effectively cope with day, night, and seasonal solar cycles through batteries and heat storage. Energy storage has yet to be widely adopted, but many new storage technologies, particularly in the form of advanced batteries, are beginning to become commercially available. Solar capture is dependent on clear weather, consistent sunshine, and placement of installations in favorable geographic locations. For almost every region of the world, however, solar technology has significant room for growth. More efficient technologies combined with lower costs promise to make solar an ever-larger part of our global energy system. The use of solar power alongside other renewable technologies such as wind or hydropower can also mitigate the problems associated with solar's variability. In many places, Colorado or Wyoming, for example, wind and solar electricity generation complement each other well. The wind blows most consistently in Colorado and Wyoming at night and during the winter, when solar is at its worst.

Solar Heating

The nature of solar energy allows us to harvest it in a variety of ways, both complex and simple. **Passive solar heating** is the simplest way we can effectively harness the sun's energy to heat buildings without any collectors or moving parts. Passive solar designs attempt to maximize **solar gain**, which is the amount of energy contained in sunlight that is absorbed as heat within a structure. For example, the Anasazi in the southwestern United States oriented their buildings to maximize solar gain (Image 9.6).

The principles of passive solar building design are relatively simple (Image 9.7). First, many south-facing windows are needed to let winter solar heat in. Next, good insulation in walls and windows help to keep the heat in. And water, masonry, or rock

Image 9.6 Prehistoric societies often built buildings to take advantage of solar heating and light, such as the cliff dwellings of Mesa Verde, Colorado.

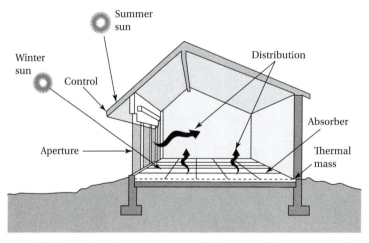

Image 9.7 A passive solar system uses sunlight to provide heat using windows and substances with a low specific heat, like tiles, to capture and radiate heat.
Office of Energy Efficiency & Renewable Energy / Wikimedia Commons / CC-BY-SA-3.0

beds help spread the sun's heat through a full day and night and sometimes longer, keeping the building from overheating during the day and getting too cold at night. It is tricky to balance these factors, plus keeping a building affordable, with the variability of the solar resource to ensure a home that stays comfortable all the time.

Passive solar heating is limited by the amount of energy that can be absorbed into a structure or tank. Mechanical systems can help us capture far more solar energy for heat in a process known as **active solar heating** (Image 9.8). These systems capture

Image 9.8 Diagram of a solar water heating system, a common type of active solar system.
Stan Zurek / Wikimedia Commons / CC-BY-SA-3.0

(a)

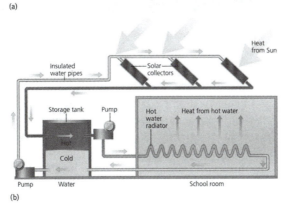

(b)

solar energy to heat spaces or water using solar collectors, pumps, or fans. Active solar systems are commonly used in many regions of the world to heat water, which can otherwise require use of a significant amount of electricity or fossil fuels. Water can be heated cheaply to low temperatures for use in swimming pools or to higher temperatures with slightly higher grade equipment for use in standard residential applications such as home heating, bathing, and cleaning.

A variety of collectors are used for active solar water heating, the most common type being the **flat plate collector** (Image 9.9). Flat plate collectors have a sheet of glass on the front, an air space, and then a flat, dark-colored solar absorber with tubes running behind it to absorb the heat. That heat is passed to water or antifreeze pumped through the tubes. The heated water can then be stored in a reservoir or tank, sometimes underground.

Evacuated tube collectors are another popular active solar technology and are more common in colder climates or where higher temperatures are needed. These systems use sealed glass tubes with a vacuum inside to increase efficiency and decrease convection loss. Finally, a heat absorber is placed in the middle and connected to water or antifreeze in tubing. Evacuated tube collectors tend to be more expensive

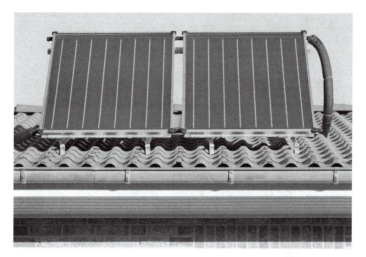

Image 9.9 Flat panel photovoltaic systems can be installed on rooftops to provide some electricity to the structure they are located on.
Bjoern Wylezich / Shutterstock

than flat-panel systems, but are more efficient in lower ambient temperatures, making them useful in colder climates. Many evacuated tube collector systems have had difficulty because of how fragile the vacuum system must be, resulting in systems with high mechanical failure rates. Calculating the efficiency of solar hot water systems can be difficult because the efficiency of each type varies depending on latitude, cloud cover, how much of the heated water is used, and how the system is installed on a building.

In addition to heating water, active solar systems can be used as air heaters, with similar mechanical principles. For instance, collectors resembling the flat plate systems can be used to allow solar energy to directly heat air, which can then be blown into a building for direct space heating. These systems are generally cost-efficient, especially where traditional heating costs are high. Additionally, the equipment involved tends to be relatively simple, requiring little building retrofitting or specialized manufacturing. Despite their benefits, solar air heat systems have not been implemented widely, with many green building projects opting for passive solar heating instead.

Photovoltaics

Solar technology has been developing rapidly in recent years, through enormous improvements in efficiency and cost and changes in policy. These advances have been especially significant in the field of solar photovoltaics(PV), which have transformed from the relatively inefficient, expensive electrical generators of just a decade or so ago to the third most prevalent form of renewable energy worldwide, after hydro and wind power.

Photovoltaic panels take advantage of the photovoltaic effect to generate electricity. In the photovoltaic effect, light energy excites electrons, encouraging them to move. The moving electrons become an electric current, which can then be transported from the panel. By optimizing materials and design, more and more of the energy in sunlight can be captured as electricity. Advances in PV technology and mass production have accelerated even in the past two to three years, causing the price of PV systems to drop precipitously, making both small- and large-scale solar generation economical in many regions.

Photovoltaic technologies are divided between **concentrating** and flat panel nonconcentrating cells (Image 9.10), with the latter being far more common at present. Conventional nonconcentrating solar cells are generally constructed out of silicon and integrate many crystalline silicon solar cells into an array, usually in the form of a solar panel. Newer technologies have become available for **thin-film photovoltaics** (TFPV), which are less efficient than conventional bulk silicon cells, but are cheaper to produce. Because TFPV cells are often literally flexible and their manufacturing process can be adapted to coat a variety of materials, they can be integrated into buildings seamlessly, allowing for building-integrated photovoltaics (BIPV), in which the panels are built into the structure of the building. In practice, this involves incorporating PV into windows, walls, and other external structures (Image 9.11). Even with their lower costs, TFPV systems are less efficient than standard PV, and therefore they remain a relatively small part of the overall solar market. The immense reduction in the cost and improvements in efficiency of traditional silicon solar cells have ensured that they remain by far the largest part of the overall solar market in

Image 9.10 Flat panel photovoltaics are the most commonly used type of solar cell.
Diyana Dimitrova / Shutterstock

Image 9.11 Concentrating photovoltaics use lenses to focus solar energy and produce larger amounts of electricity.

Dennis Schroeder/Science Photo Library

the United States. Thin-film and building-integrated PV systems will likely be used in specialized roles as their technologies improve.

While conventional silicon solar panels have dominated the rapidly expanding solar market in recent years, CPV cells have demonstrated great potential in terms of efficiency, which has so far been offset by their much higher manufacturing and installation prices and more specific siting demands. Similar to flat panel PV cells, CPV cells directly convert sunlight into electricity, but they also include a focusing array of lenses and/or mirrors that physically concentrate solar radiation onto the cell. For certain PV materials, this increases the amount of energy that can be captured for electricity.

While standard silicon cells average between 14 and 19 percent efficiency among commercially available cells, some experimental CPV cells have demonstrated over 40 percent efficiency. Despite their technical advantages, CPVs have several practical disadvantages that limit their application. The optics used in CPV arrays cannot concentrate diffuse light. Slight haze seriously degrades CPV output as the light is scattered. Concentrating PV systems need high, dry desert environments—high because less atmosphere means less scattered light and dry because even moisture in the air scatters light. Flat panel PV can be installed in far more parts of the world than CPV.

Further, CPV systems focus a large amount of radiation onto a small area, meaning the cells absorb a lot of heat. These high amounts of absorbed heat require advanced cooling systems, which increase the size and cost of manufacturing and installation. As technology improves, CPV cells are likely to become more common in certain circumstances and applications. The massive drops in price of conventional silicon cells and TFPV systems are likely to keep them in the lead for most uses; however, CPV systems may still prove to be effective in certain circumstances (Table 9.1).

Table 9.1 Solar Cell Types

	Advantages	*Disadvantages*
Flat panel (conventional bulk silicon)	Relatively low cost for high output, most common current type	Less efficient than concentrating systems
Thin film/building integrated	Cheapest to construct, most adaptable for construction	Significantly lower efficiency
Concentrating	Most efficient technology by a large margin	Much more expensive than other systems, requires more select locations, can present infrastructure challenges

Solar Thermal Engines and Electricity

Unlike PV solar power generation, solar thermal power, known as concentrating solar power, uses the heat of the sun to power a thermal engine that generates electricity. Before the massive drops in cost and advances with PV technologies, CSP had been a primary focus for solar power development and at one point led in alternative energy investment. Photovoltaics have significantly overtaken CSP in market share, yet new opportunities for solar thermal applications are still being pursued, particularly in places where climate and geography are optimal for CSP, such as desert regions (Image 9.12). Additionally, CSP plants offer distinct advantages over PV technologies that might make them desirable in our energy basket.

Where PV technologies take advantage of the PV effect to convert solar radiation directly into electricity, CSP technologies concentrate solar heat to high temperatures to make steam to power thermal turbines. In terms of its implementation as a major energy resource, CSP has much more in common with thermal-based fossil fuel generation than it does with solar PV. To optimize operational efficiency, CSP plants are generally large and are often far from the cities where we use most of our electricity. Photovoltaic systems can be small home-sized units or large utility-scale installations. Small PV systems can be distributed close to or on top of our homes and businesses. Concentrating solar power plants, because of their large size and reliance on electricity transmission, are deployed in a manner more like traditional thermal power plants. Further, some CSP plants store heat for generating electricity at night. This makes them inherently less variable than PV technologies. Despite these advantages, CSP has yet to achieve a level of efficiency and cost that is conducive to widespread use; this has allowed PV installation to skyrocket, where CSP use worldwide is still limited and, in many places, only experimental. Additionally, CSP plants are only viable under certain geographic conditions, limiting their use to generally sunny regions with already-developed electricity infrastructure.

The most common type of CSP plant, both historically and at the current time, is the **parabolic trough** system (Image 9.13). A solar parabolic trough generator uses a curved

Image 9.12 This collector dish focuses solar radiation and then uses the heat to power a conventional thermal electricity–generating system.
Iaian Frazer / Shutterstock

Image 9.13 A parabolic trough solar plant in Spain.
Ashley Cooper/Science Photo Library

mirror to concentrate sunlight on a central tube, generally containing a type of thermal oil, which is heated by the solar energy and piped out of the system (Image 9.14). The fluid, heated to around 400°C (750°F), is used to produce steam for electricity generation or occasionally to power machinery. To produce power during dark hours, some parabolic trough plants are augmented with other fossil fuel–based generators (usually natural gas), which must be limited to a quarter or less of the total generation to

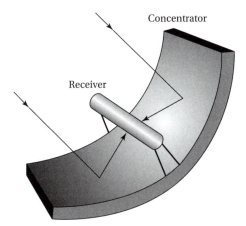

Concentrator

Receiver

Image 9.14 Parabolic trough systems concentrate heat along tubes that contain superheated fluids that can then be used to run a traditional thermal power plant. Courtesy of the Solar Energy Development Programmatic EIS

qualify as a renewable energy plant by the U.S. Department of Energy. Some more recent trough plants have experimented with heat storage systems that can be used for overnight generation from daylight-accumulated heat using molten salt. The efficiency of solar trough systems, around 15 percent, is comparable to that of many PV systems. Nonetheless, trough systems have not kept pace with the cost improvements seen in PV systems and continue to require larger installations.

Parabolic trough systems have been primarily installed in the southwestern United States and Spain, both of which are generally considered the leaders for all types of solar thermal power plants. The largest trough system is the Solana Generating Station in southern Arizona, with an installed capacity of 280 MW, larger than most other plants of similar design. The Solana facility additionally uses molten salt to store heat to generate solar electricity overnight, mitigating some of the variability concerns associated with solar PV. Some trough plants, like, the Solana facility, require large inputs of water for cooling. This is a particular concern in the arid environments that are most conducive to solar trough installations.

Stirling dish systems are another kind of CSP that use mirrors to reflect sunlight on a receiver, but instead of using long troughs to heat fluid that is subsequently piped out of the heater, they focus heat energy on a single receiver that contains the generating unit itself (Image 9.15). Because a greater amount of light can be focused on the source, dish systems create higher temperatures (around 750°C or 1380°F), which can be used to power a Stirling engine, rather than a steam turbine. Stirling engines, invented in the early 1800s as an alternative to steam engines, are more efficient than other CSP technologies (Image 9.16). Dish systems can approach 30 percent efficiency, making them one of the most efficient solar technologies of any kind.

While dish systems are efficient, they suffer from problems that make them less attractive than other solar technologies. Because the dish itself must be directly pointed at the sun to properly concentrate energy, the installation requires a multiaxis tracking system, whereas trough systems only need to track in one direction. The dishes can be large and heavy, making these tracking systems expensive. Additionally, the greater number of mechanical parts in the Stirling engines necessitates more maintenance,

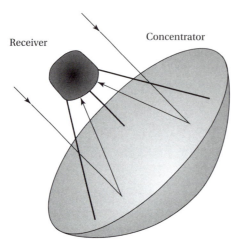

Receiver

Concentrator

Image 9.15 This diagram shows how a Stirling system generates electricity.

Courtesy of the Solar Energy Development Programmatic EIS

Image 9.16 Stirling dish systems have only been used experimentally.
Bill Timmerman/Science Photo Library

especially when compared to PV systems, which have no moving parts. These costs and concerns have meant that few dish systems have been built, and some that have been built have since been dismantled.

A promising technology that has been employed in limited areas is the **solar power tower**, which uses a field of mirrors focused on a central heat absorption source, which then generates power from a steam turbine (Image 9.17). As with some CSP solar heat storage systems, the power tower uses molten salt to transfer heat. As with CSP systems, molten salt can be stored, allowing generation after the sun sets. Like other CSP plants, power towers take up large amounts of land, must be installed in dry desert areas, and can be environmentally controversial in the fragile desert ecosystems that are preferred for CSP (Image 9.18).

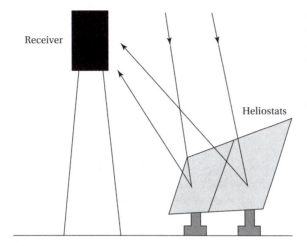

Image 9.17 Solar power towers use vast arrays of mirrors to focus solar energy on the top of the tower and use molten salt to heat a thermal power plant.

Courtesy of the Solar Energy Development Programmatic EIS

Image 9.18 Several solar power towers have been built, such as this facility in Seville, Spain.
Quintanilla / Shutterstock

Some power tower prototypes cover multiple square kilometers, making their ecological impact especially controversial.

The original testing of power towers took place through the Solar Project, which consisted of two different tower designs in the 1980s, Solar One and Solar Two, in the Mojave Desert in California. These plants, which produced 10 MW each, have since been decommissioned and bulldozed, but the concept was successful, leading to the planning and construction of several other tower plants in other locations. The largest single generator is currently the Ivanpah facility, also located in the Mojave

Desert. The plant uses several tower systems to generate a maximum capacity of 392 MW, less than an average coal-fired power plant, but significantly greater than most solar plants. Several of the current tower plants are testing technology and engineering solutions that aim to further drive down the costs of tower generating, and the U.S. Department of Energy estimates that power tower electricity could be cost-competitive by 2020.

Other CSP technologies are also in development, with the possibility of scaling up to commercial use in the future. Fresnel reflectors, which are similar in size and concept to trough systems, may offer increased efficiency while being economical as smaller projects and are in testing in several locations. Further, CSP systems could potentially be modularly built in small sizes and used in a way similar to PV systems for distributed or remote location generation. The increasing efficiency and decreasing cost of PV systems makes small CSP systems unlikely to be competitive for the foreseeable future, however.

Concentrating solar power was the major target of solar technology investment and development in previous decades, resulting in major projects like Solar One. The past several years have seen massive advancements in the technology, costs, and efficiency of PV, while CSP technologies have stayed relatively constant. Consequently, PV installations have skyrocketed while CSP has not. Aside from plants in the southwestern United States and some research plants in Spain, CSP appears unlikely to experience growth on the same scale that PV has enjoyed recently. Because of its continuing high costs and the fact that some PV technologies are approaching CSP in efficiency (its major initial appeal), CSP will likely experience lower global implementation in the coming years, barring major technological or economic changes.

Worldwide Solar Generation

Worldwide, solar power has seen significant growth over the past decade. In response to signing the Kyoto Protocols, several European Union countries, led by Germany, Italy, and Spain, have dramatically increased their solar power generation as part of an effort to curb greenhouse gas emissions (Table 9.2). Recent growth in the United States and China is also playing a key role in expanding the use of solar power around the globe, and more proposed systems are being planned in the Middle East and North Africa. Additionally, in remote locations and less developed countries, solar is beginning to be seen as a way to help bring small amounts of electric generation to even the most far-flung populations.

It is surprising to note that much of the solar power currently produced around the world comes from places that are relatively rainy and far from the equator, thus in areas that have limited solar potential. Most of the leading solar power producers are in Europe, including countries like Germany, which is known for its relative lack of sunny days, short winter days, and wet climate. These areas have a low solar potential, meaning that the total amount of solar energy that reaches the earth's surface is low. The world's highest solar potentials are found in desert regions closer to the equator, such as the Sahara Desert, the Arabian Peninsula, or Australia. Apart from Australia, none of these regions has major solar projects in place at the scale seen in places like Germany, Spain, or Japan (Image 9.19).

Table 9.2 Top Ten World Solar Photovoltaic Electricity Generators by Installed Capacity (in gigawatts), 2016

China	78.1
Japan	42.8
Germany	41.2
United States	40.3
Italy	19.3
United Kingdom	11.6
India	9
France	7.1
Australia	5.9
Spain	5.5
World	303

Source: IEA-PVPS, 2017, 2016 Snapshot of Global Photovoltaic Markets, http://www.iea-pvps.org/fileadmin/dam/public/report/statistics/IEA-PVPS_-_A_Snapshot_of_Global_PV_-_1992-2016__1_.pdf

Image 9.19 Spain has installed a vast amount of solar power facilities, including many photovoltaic panels like these at an orange farm.
Philip Lange / Shutterstock

Despite the relative dearth of large-scale solar projects in many parts of the world, use of solar energy to heat water is commonplace in developing countries (Image 9.20). Black rooftop water tanks were pioneered in Israel in the mid-twentieth century and are now ubiquitous in many developing countries with warm climates, from Turkey to the Caribbean to Peru. Many traditional housing designs also utilize solar energy for heat and lighting.

Future solar growth is likely to be high in desert regions across the world because of increasing interest in alternative energy, even within the major oil-producing countries of the Middle East. The International Renewable Energy Agency is based in the United Arab Emirates and aims to increase solar power research through international cooperation. Other wealthy Arab states, including Saudi Arabia and Qatar, have also made commitments to drastically increase solar power as part of their energy systems, including for saltwater desalinization. In poorer developing countries, PV systems are touted as a means to produce electricity in off-grid locations. In small isolated villages, small-scale PV systems are starting to appear that power cell phone towers and refrigeration systems, among other applications. As PV prices fall, many off-grid areas may turn to the technology to produce some electricity.

Image 9.20 Solar water heating systems are widespread in hot and sunny climates, like this system on a roof in Vietnam.

Germany's Solar Revolution

Germany is one of the world's leading countries in solar electricity generation, despite its unfavorable geography compared to much of the rest of the world (Image 9.21). Germany has a cool, cloudy climate and receives far less sunlight than most of the continental United States, for example, making the equivalent output of any solar panels lower than it would be elsewhere in a given year. The German government's commitment to solar energy has overcome much of this disadvantage and allowed the country to be at one point the world's largest generator of solar electricity, only surpassed as of 2017 by China and Japan. Germany's solar network is almost entirely PV because solar thermal generation is extremely inefficient with the lower levels of solar exposure in the country.

The massive growth of solar electricity in Germany has largely been driven by public policy (Image 9.22). The government has set minimum targets for renewable electricity, aiming to generate 35 percent of the national total from renewable sources by 2020, increasing to 80 percent in 2050. To distribute generation effectively and ameliorate variability, Germany has focused on a combination of solar electricity and wind generation, which now account for almost 16 percent of total electric generation together, while solar's share of the total is 6.2 percent.

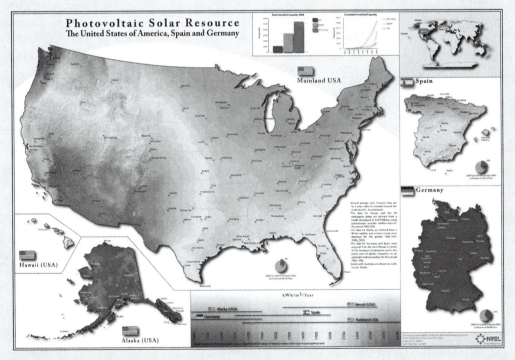

Image 9.21 The United States has more solar potential than many countries with more vast solar systems, like Germany and Spain.

Courtesy of the National Renewable Energy Laboratory

continued

continued

Image 9.22 Solar panels are found on buildings all over Germany, such as the solar panels on these farm buildings in rural Germany.
Hennadii Filchakov / Shutterstock

While Germany's total solar output (41.2 GW installed capacity in 2016) is impressive, growth of solar generation has slowed considerably. Large subsidies, particularly in the form of feed-in tariffs, led to strong growth from 2008 through 2012, a period in which almost 27 GW of capacity was installed. Federal subsidies for solar generation have been cut significantly since 2013, leading to less than 2 GW of capacity being installed in 2014, and only an additional 3 GW in the following two years. The government aims to keep subsidies at a level that guarantees steady growth in renewable electricity, but annual growth after the subsidy cuts will be significantly smaller than during the boom period. Strong growth in China and Japan has removed Germany from its leading global position, though it remains a major player. Even so, its continuing growth in solar generation and outsized national total demonstrate that strong government support and a receptive public can drive impressive developments in solar electricity, even in a country with a relatively unfavorable geography.

Solar Generation in the United States

Solar energy generation in the United States has grown rapidly in recent years, largely a result of improvements in the affordability of PV. Solar power accounts for only about 1 percent of total electricity generation in the United States at present, but its growth puts the country on target to be one of the larger sources of renewable energy by 2040. The growth in the United States has been significantly helped by the large number of solar research companies located there, as well as the availability of tax incentives in certain states. Because of the decentralized nature of energy policy in the United States, progress has been much faster in some places than in others, and the additional variation in geographic conditions for solar means that greater increases can be expected in sunnier regions.

California leads the country in solar projects, having set a goal of one half of all energy coming from renewables by 2030. Major solar power plants have largely been constructed in the southwestern United States because of the higher amount of sunlight received in that region and the large amounts of open land needed for implementation (Images 9.23 and 9.24). Major plants have been built or are in development in California, Nevada, and Arizona and several other states have developed solar power on a smaller scale. Installations of distributed PV have been rising dramatically as well, driving much of the growth in solar. Because distributed generation is harder to accurately track, the installed capacity of PV is a rougher estimation than that of other energy sources, but has grown significantly in recent years.

Because of the drastic reductions in cost for solar power, as well as the significant policy incentives that exist in certain states, the pace of growth for new installations is forecasted to continue at its current rate. Almost 5 GW of large-scale capacity is currently under construction in the United States, and over 30 GW of capacity is in the planning stage, all of which would nearly double the current supply of solar energy. These numbers do not account for the continuing growth of distributed solar, which will be a major part of the solar energy generation capacity in the future as many private homeowners and businesses continue to install PV systems on their property. Solar energy still represents a small part of the total grid in the United States, but the advances in technology and policy have paved the way for it to grow into a major component of the country's overall energy supply in the future.

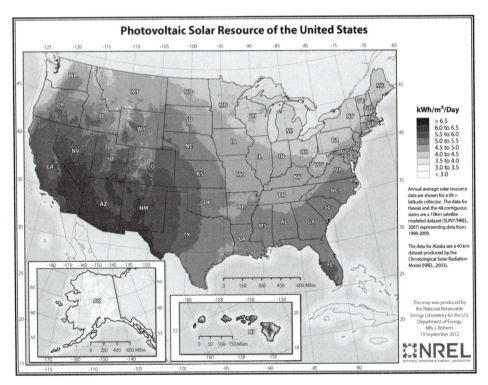

Image 9.23 Photovoltaics are viable for widespread use across the United States.
Courtesy of the National Renewable Energy Laboratory

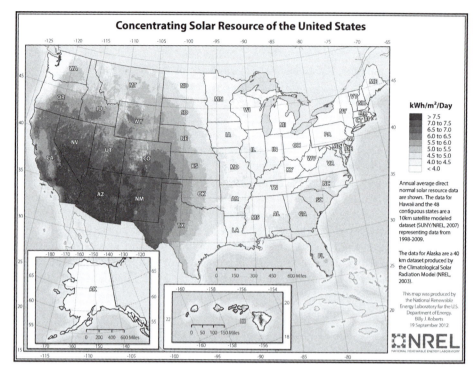

Image 9.24 Concentrated solar power systems are best built in the southwestern United States, rather than in other less sunny areas.

Courtesy of the National Renewable Energy Laboratory

Environmental Impacts

Solar power is widely seen as one of the most environmentally friendly energy sources because its very nature as a renewable precludes significant greenhouse gas emissions or intensive resource consumption. Despite these advantages, solar does have a variety of environmental impacts that must be assessed in comparison to the systems it is meant to replace.

The generation process for solar electricity, both from PV and from thermal sources, does not emit any particulates or gases (with the exception of thermal plants that use natural gas augmentation). The production of solar generation equipment, however, can entail the use of materials that are environmentally harmful to mine, smelt, or otherwise produce, and the production process of the systems, especially PV panels, can involve measurable emissions impacts. As panel technology has developed, the impacts and emissions from production have changed greatly, making a basic comparison difficult. With complete life-cycle greenhouse gas emissions of the most efficient combined cycle natural gas plants being about 400 grams per kilowatt-hour, solar technologies are clean, coming in at between 20 and 50 grams per kilowatt-hour. Future scaling

VIGNETTE
Large-Scale Solar in the Mojave Desert

Many regions of the United States have favorable geography for solar power. The American Southwest, famed for its deserts, is by far the richest in solar potential and has consequently been the site of the largest solar energy projects in the country. The Mojave Desert, which occupies much of Southern California and Western Nevada, is ideal for large solar projects for two reasons: its strong solar resource and its proximity to some of the largest urban areas in the country.

The Mojave Desert has been the center of major solar developments since the construction of Solar One, one of the first major CSP plants, in 1982. Since then, several large projects have been constructed, with a total installed capacity of 1,459 MW by the end of 2014. The Mojave is unique in that a major share of the large plants are solar thermal systems (1,060 MW vs. 994 MW of PV), which means that much of the solar energy from the area can continue generating at night with stored heat. The proximity to Los Angeles, Las Vegas, and San Diego gives solar projects in the Mojave a massive electricity market with good transmission infrastructure. Consequently, there is still a great deal of room for growth in solar power in the region. The largest plant in the region, the Ivanpah Solar Facility, opened with a capacity of 392 MW, making it the largest single solar facility in the world (Image 9.25). New projects as large as 500 MW have been planned, with several gigawatts in solar power projected to be constructed in the next several years.

Not all growth in solar generation in the Mojave has been welcome, however. The Soda Mountain solar project was planned to be built in Riverside County with a capacity of 358 MW, but has encountered strong resistance from environmental groups. Soda Mountain is close to the Mojave National Preserve and has large populations of bighorn sheep and desert tortoises, which environmental groups have said would be threatened by intensive solar development on the site. Bechtel, the site's developer, has worked with the federal government to reduce the size of the plant to 264 MW to lower the impact on wildlife, but the city of Los Angeles has declined to purchase power from the project, preferring instead to get solar electricity from cheaper and less invasive sources. Despite opposition to certain solar projects, many areas of the Mojave are ideal for continuing solar development, and strong growth in solar generation is likely to continue in the area for the foreseeable future.

Image 9.25 The Ivanpah solar plant is located in the Mojave Desert.
Piotr Zajda / Shutterstock

of production and improved production processes have the potential to lower these emissions figures even further. When compared to coal and other fossil fuel–generation technologies, solar is especially clean, having lower emissions by orders of magnitude (coal can produce as much as 1,000 grams of greenhouse gas per kilowatt-hour). In terms of emissions, solar energy is clean, especially when considered over the lifetime of the technology and not just its production.

Additionally, large-scale projects involve large land use. In certain cases, this can infringe on the habitats of endangered or rare species and provoke land use disputes. In the United States, this has played out in some of the projects in the desert southwest, where solar sites have been opposed by groups trying to protect desert tortoises and spaces that hold religious importance for Native American communities. In recent tests, there have also been concerns over birds being burned alive at solar power tower sites.

REVIEW QUESTIONS

1. What can passive solar power be used for?
2. What differentiates passive and active solar power?
3. Why is PV overtaking CSP in global deployment of solar power?
4. How can the lack of sunlight at night be overcome with solar power? Does this work equally with PV and CSP?
5. Why does Europe lead in solar development, despite its relative lack of sunlight compared to other regions?
6. Why are deserts the best place for solar power? Does altitude matter too?
7. Which U.S. states are likely best for solar and worst? Why?

Image 10.1 Stuart Monk / Shutterstock

CHAPTER 10

Biofuels

Biofuel Past and Present

Biofuel is the longest-used energy resource in human history, with wood and peat fires being among the first advances of early hominids—even *Homo erectus* (an early ancestor of the human species) harnessed fire. Heating, cooking, and some simple industries remain large consumers of wood-based biofuel in many parts of the world. A **biofuel** is any recently grown organic material that can be used for its energy, usually through combustion. For much of history, and today in less developed countries that lack expanded or stable power grids, biomass in the form of wood, animal dung, or plant waste has been burned for cooking and heating as a primary source of energy (Image 10.2). Wood biomass was the main historical source of energy worldwide and was used for almost all purposes until the Industrial Revolution. Until coal mining became sufficiently large in scale, wood or other biomass such as peat was effectively the only available combustible fuel source.

Developments of newer technologies such as steam turbines allowed biological products to be used to produce electricity in **waste-to-energy** (WtE) plants during the twentieth century, reducing waste disposal problems, with possibly some increase in air pollution, while generating electricity. While WtE is not a major source of electrical generation worldwide, it remains widespread enough to contribute significantly at

Image 10.2 Biomass, in the form of wood or dung, has always been a primary fuel for humans. Wood-fired cooking stoves remain common in poor parts of the world.
WeStudio / Shutterstock

a local level in many locations. Later technological advances in refining, fermenting, and biological processing allowed for the production of transportation fuels including ethanol from biological products beginning in the 1970s. In large agricultural countries such as Brazil and the United States, ethanol is a major source of energy, either as a fuel additive or as a nearly pure fuel in itself, with only a small percentage of the energy in the ethanol coming from fossil fuels.

As an energy resource, basic biomass is problematic in many ways. While it is renewable, given that trees and plants can be regrown, population growth in most places that rely on biomass has meant that forests are cleared far more quickly than they can regrow. Consequently, local wood resources cease to be renewable in practice. In some of the poorest countries, Haiti, for example, deforestation has been so pronounced that wood has become nationally scarce (Image 10.3). Additionally, the burning of old-growth forests releases large amounts of soot and carbon dioxide (CO_2), removes valuable natural protection against erosion, and exposes local water supplies to higher degrees of contamination. And deforestation often leads to desertification—or once-productive areas changing into deserts. While much of the current use of biomass involves simply combusting wood or other organic materials for heat or power, many developing technologies aim to produce transportation fuels from agricultural or wood waste, effectively harnessing the energy in organic materials that would otherwise be burned or disposed of. Much of the future of these advanced transportation biofuels depends on continued research, as well as the development of such competing technologies as electric cars.

Broadly speaking, biofuel can be derived from many different types of sources for end use in a wide range of applications. Biofuels can be combusted to generate electricity, either as agricultural waste to power ethanol production facilities or to supply municipal grids from WtE or biomass facilities. More recently, biofuels have been

Image 10.3 Cutting down trees for firewood can have devastating consequences when overdone. The border between Haiti, where a large poor population has been forced to deforest large areas, and the neighboring Dominican Republic, which is better off economically and less reliant on firewood, is shown here from space.
James Blair / National Geographic Creative

developed for use as transportation fuel, in which the biological source is converted into an energy-dense liquid such as ethanol or biodiesel for direct use in vehicles. This sector has seen immense growth in recent years as society has attempted to reduce vehicular pollution and dependence on fossil fuels, which have high volatility on the global market. Because the number of current and possible biofuels is large and because these fuels can be produced in many ways with varying efficiency, it can be difficult to broadly generalize about biofuels. Nonetheless, the world biofuel market has grown rapidly, and new technological advances suggest that biofuels will continue to grow in importance to the global energy economy.

Transportation Biofuels

To reduce the dependence of cars, trains, ships, and aircraft on fossil fuels, substantial research and investment has been devoted to the development of liquid biofuels for transportation. The largest of these has been through **energy crops**. Rather than growing traditional food products for human consumption, energy crops take agricultural products like corn, soy, or sugarcane and process their natural energy content into fuels such as biodiesel or ethanol. In principle, energy crops can reduce carbon emissions because their carbon is recently sequestered from the atmosphere, while carbon from fossil fuels has been fixed for millions of years. In practice, energy crops like corn and soy must compete with food production for land and resources, which has made them highly controversial. Additionally, the actual energy yield of energy crops is hard to pin down, given the large number of energy-intensive inputs, such as fossil fuels to run farm equipment, fertilizers usually made from natural gas, and the energy needed to pump water for irrigation. Further issues related to converting more land area into farmland (thereby destroying natural land cover in the process), increased fertilizer pollution of rivers and groundwater, and excessive or unsustainable water use complicate the development of energy crops. Technological advances that could process agricultural waste into fuel or potentially use low-impact crops such as switchgrass or the residue of food crops as a fuel source may be able to mitigate these concerns in the long term. So far, these more sustainable fuels have yet to become economical to produce, primarily because we do not have the technology to easily convert these products into liquid fuels.

Ethanol

Energy crops can be used to produce multiple fuel types, but **ethanol** is the most common form of agricultural biofuel worldwide. Ethanol (also known as ethyl alcohol) can be produced from almost any organic material through various chemical processes, although sugarcane and corn are currently the largest sources (Image 10.4). Ethanol use has grown rapidly in the United States and Brazil, which both possess large quantities of arable land and extensively developed agricultural infrastructure. Other countries have begun to invest in ethanol production technologies, but the world market is still primarily dominated by American **corn ethanol** and Brazilian **sugarcane ethanol**.

Image 10.4 Sugarcane is the most efficient crop for producing ethanol.
Thomas Amler / Shutterstock

Cellulosic ethanol, made from nonfood crops such as switchgrass or from agricultural and organic waste, may be a more sustainable approach to ethanol production in the long term. Research into several types of cellulosic ethanol is ongoing, primarily focusing on developing enzymes that can turn these biological feedstocks into alcohol. We are just starting to see cost-competitive, commercial-scale cellulosic ethanol plants.

The earliest ethanol energy crop to be developed on a large scale was sugarcane, primarily in Brazil. The 1973 world energy crisis caused many governments to examine their dependence on gasoline imports, and Brazil's large production of sugarcane (with large amounts of land available to expand into) showed a unique opportunity to experiment with using agriculturally derived ethanol as a major fuel source. Sugarcane is attractive as a fuel crop because its 20 percent sugar content by weight gives it a high ethanol yield, especially compared to ethanol made from corn, the dominant feedstock for biofuel in the United States. Despite this efficiency, sugarcane requires large amounts of agricultural land for cultivation. If this land is cleared from previously forested areas, the fixed carbon in the trees will be released, and the net impact of sugarcane ethanol in reducing atmospheric carbon will be lessened. Despite this, several studies have calculated that the CO_2 emissions impact of sugarcane ethanol is still positive, even accounting for land use changes. Because sugarcane is not a primary food source, the increases in sugarcane ethanol production will not cause food prices to rise, making it safe as an energy crop. Sugarcane ethanol has been able to compete without direct subsidies in Brazil, although much of this has to do with the high taxes on gasoline meant to encourage ethanol use. Because of government encouragement, the majority of commercial vehicles use ethanol, and now 50 percent of all Brazilian fuel consumption for vehicles comes from ethanol.

Image 10.5 There are many ecological concerns with the ethanol industry in Brazil, such as further deforestation of the Amazon River basin.

Frontpage / Shutterstock

Sugarcane ethanol is produced by harvesting the cane, processing it into a sugar-laden syrup, and then fermenting that syrup into an alcoholic liquid. The resultant liquid, which is about 10 percent alcohol, is then processed to create nearly pure ethanol. The drier waste products of the sugarcane harvesting, the *bagasse*, are generally burned on-site to provide the electricity and heat needed to power the ethanol production, with excess electricity being sold to the grid. Given the WtE generation and the relatively efficient process of fermentation, sugarcane ethanol production is economical and low impact. Improvements in plant technology, along with the breeding of varieties of sugarcane more suited to ethanol production, have led to steady improvements in productivity and sustainability within the Brazilian ethanol industry, much of which is responsible for the comparatively large reduction in greenhouse gases (GHG) that ethanol use has allowed.

Corn Ethanol

The biofuel industry in the United States, while partially modeled on that of Brazil, has developed differently. Much as Brazil was inspired to develop a domestic agricultural fuel industry by the high oil prices of the early 1970s, the major impetus for the development of the U.S. ethanol industry was the rapid growth in oil prices of the mid-2000s. While Brazil chose sugarcane as the primary source of ethanol, the overwhelming majority of ethanol in the United States is derived from corn (Image 10.6). Corn has a much lower sugar content than sugarcane and is more difficult to process into ethanol, making its average energy yield lower. The climate of much of the United States is not conducive to growing sugarcane, so another major feedstock for ethanol had to be found. The dominant position of corn as the major crop in the U.S. midwestern agricultural heartland, combined with significant production subsidies because many major corn-producing states are politically powerful, made corn the primary U.S. source of ethanol, a position it has maintained.

The overall energy balance of corn ethanol is still under debate; consequently, the promotion of corn ethanol as an environmental alternative to gasoline remains controversial. In comparison with sugarcane ethanol (the only other commercially significant type currently in production), corn makes less ethanol, is less efficient in land use, and is more energy intensive to produce (Image 10.7). By weight, sugarcane contains about twice the sugar of corn and is easier to refine. The bagasse from sugarcane production can completely supply the energy for the refining process, further improving the energy balance of the crop. Sugarcane also uses less fertilizer than corn per acre. According to research by the U.S. Department of Energy, corn has a barely positive energy balance, yielding about 25 percent more energy in ethanol than is used to produce it (compared to around an eightfold gain for sugarcane), but is also water and fertilizer intensive. One gallon of corn ethanol consumes several hundred gallons of water for production (depending on where the corn is grown and how it is processed). In drier areas that rely on aquifers and less stable water sources, this massive water use can be problematic. Further, the large amounts of fertilizer used in corn production exacerbate water pollution and runoff issues.

Image 10.6 Corn is the primary crop for producing ethanol in the United States.
Dan Thornberg / Shutterstock

Image 10.7 Corn is taken to this processing plant to be converted into ethanol.
Jim Parkin / Shutterstock

As a major food source, particularly for lower income populations across Central and South America and parts of Africa, the increased use of corn to make fuel instead of food has contributed to several significant corn price spikes, which have reduced the food security of many populations. The debate between energy security and international food security with energy crops is discussed further in a vignette at the end of this chapter.

The production process of corn ethanol is similar to that of sugarcane ethanol, with a few notable differences. In the most common type of processing, known as dry milling, the corn kernels are ground and then added to a liquid that is heated and fermented with enzymes for about two days. The enzymes and heat convert the corn starches into sugar, which is then fermented directly into alcohol. After the alcohol is separated, the remaining pulp is rendered into a concentrated livestock feed. Carbon dioxide is often captured from the fermentation step and is used in industrial and food applications. While the ethanol production process cannot be powered by its own waste, as with sugarcane, the byproducts of agricultural feed and CO_2 significantly offset production costs, as well as recycling what would otherwise be waste products.

Cellulosic Ethanol

The types of ethanol produced from crops, known as first-generation biofuels, have many drawbacks relating to efficiency, food prices, and land use changes (Table 10.1). Second-generation biofuels may mitigate most of these problems, with **cellulosic ethanol** being the most intensely studied. Cellulose and hemicellulose are polysaccharides that exist in plants, primarily in the more fibrous parts that are disposed of as waste (such as corn stalks, wood chips, or sugarcane bagasse) (Image 10.8). These compounds can be fermented and converted into ethanol, but must first be separated from lignin, the other main component in cellulose-rich parts of plants.

Table 10.1 Comparing Ethanol Types

	Corn ethanol	*Sugarcane ethanol*	*Cellulosic ethanol*
Greenhouse gas reduction compared to gasoline[a]	21% according to the Environmental Protection Agency; other sources disagree	60% according to the Environmental Protection Agency; other sources disagree	85% according to Argonne National Laboratory
Strengths	Already established as a major crop, produces useful byproducts	Most energy-efficient fuel crop, bagasse waste can be used in the refining process	Uses major waste products, does not require more land for farming, by far the most environmentally friendly ethanol biofuel
Weaknesses	Not efficient at creating fuel, may worsen emissions when including land use changes, problematic for food security, water inefficient, increases river pollution	Can only be produced in hot climates, may deplete soil quickly	Technology has not made it cost-effective yet; cannot compete with cheaper biofuels, although this may be changing

[a]Land use changes can be calculated differently, and these calculations can drastically alter these numbers.

Image 10.8 The leftovers after sugar is extracted from cane is called bagasse. This can be burned to provide heat needed for the fermentation process.
Hywit Dimyadi / Shutterstock

To make cellulosic ethanol, the cellulose must be separated from the lignin and broken down into sugars. These parts of plants have evolved to be extremely sturdy and resistant to decomposition because they are the cells that hold up the plant, which makes the refining process more difficult. Enzymatic methods, which allow different enzymes and bacteria to break down the lignocellulose into its different parts, are widely used. Thermochemical methods can also be used, which essentially cook the lignocellulose into a synthetic gas that can be converted into ethanol or another type of fuel. So far, no method of producing cellulosic ethanol has achieved commercial viability; the cost of enzymes, the larger and more expensive factories that are required, and the large energy and chemical inputs have combined to make cellulosic ethanol difficult to scale up to commercial production. Genetically engineered bacteria, commercial-scale plants, and other technological breakthroughs are reducing cellulosic ethanol production costs, but at present, the fuel represents a fraction of total biofuel production worldwide.

If cellulosic ethanol production can be advanced to the point where it is economically competitive, it would represent a major advance over energy crop biofuels. Concerns about food security, water usage, fertilizer pollution, and land use change all combine to make crop biofuels problematic. Cellulosic biofuels, however, are derived from plant waste (from corn, sugarcane, wood pulp, or other agricultural waste) or fast-growing cellulose-rich plants like switchgrass (Image 10.9). Waste-derived ethanol has the obvious advantage of using a resource that is already widely produced and generally disposed of, giving it virtually no environmental impact. Cellulose-rich plants such as switchgrass use far less water, land, and energy to grow than corn or sugarcane. Some cellulosic crops are harvested by mowing the top growth, leaving the still-living roots in place. This aids agriculture by preventing erosion and preserving soil. These crops can also be produced on marginal lands not otherwise suited for food production. Many

Image 10.9 Switchgrass grows easily on marginal land with minimal irrigation. If a process to turn it into ethanol can be developed, it could help replace food crops as a major ethanol source.
Hjochen / Shutterstock

plants considered for cellulosic ethanol grow quickly and can be harvested multiple times per year, are drought resistant, and can be grown in a variety of climates.

Cost is the primary factor preventing cellulosic ethanol from gaining widespread traction. Most cellulosic ethanol is produced either to comply with U.S. government regulations requiring a minimum use in the fuel supply or in exploratory research facilities. A further factor hurting cellulosic ethanol is the already large supply of crop ethanol; many economists believe that the U.S. ethanol industry is already satisfying the entire demand with corn ethanol, so a high-cost alternative will have a difficult task in displacing it. Increased competition from other energy sources like compressed natural gas or battery–electric cars puts further pressure on cellulosic ethanol. The many complicating factors involved make it difficult to forecast the future use of cellulosic ethanol, but it appears to have only a small role in the alternative fuel market in the short term.

Biodiesel

In addition to the various types of ethanol, one other common transportation biofuel is in use across the world: **biodiesel**. Biodiesel can be made from a wide variety of vegetable oils and animal fats and can be run in standard diesel-powered internal combustion engines. Much like corn ethanol in the United States, the majority of biodiesel is used as a fuel additive rather than a pure fuel itself (with formulations of 5–20 percent being most common), although it can also be used to power specially modified engines in its 100 percent pure form. The European Union has adopted biodiesel more widely than anywhere else, largely because of the widespread use of diesel automobiles there, although the U.S. biodiesel market has grown in tandem with the ethanol market in response to government attempts to reduce fossil fuel use.

Biodiesel can be produced from many sources, including raw plant oil, waste oil from food production, and animal fat. The lipids (fat compounds) in these oils are reacted with alcohol in a process called **transesterification**, which yields an end product known as a fatty acid ester that can be directly used in diesel engines. In the United States, half of all production of biodiesel comes from refined plant oils, especially soybean oil and rapeseed (canola) oil (Image 10.10). Other feedstocks are commonly in use for smaller scale production, including used frying oil from restaurants, recycled vegetable oil, and the waste oil from fish used to make omega-3 supplements. The great versatility in feedstock is a primary strength of biodiesel. Despite these advantages, biodiesel has several obstacles to larger scale use.

Much as with the various types of ethanol, the energy balance of biodiesel varies depending on the feedstock. When produced with waste oils, biodiesel recycles a waste product that would be produced anyway, making it sustainable. When produced directly from crops like soy, many sources estimate that biodiesel may *increase* GHG emissions because feedstock plants are often highly fuel-, fertilizer-, and water-inefficient. Scaling up biodiesel production to meet all demand for diesel fuel would require vast increases in the amount of land under cultivation and could impact food prices in much the same way as corn ethanol has. Much of the biodiesel demand in Europe is being satisfied through **palm oil** cultivation in Indonesia and Malaysia, which has led to high

Image 10.10 This refinery turns rapeseed (canola) into biodiesel fuel, which can be used in a conventional diesel engine.

TebNad / Shutterstock

rates of deforestation of fragile tropical rainforest environments. Including the land use changes, destruction of the environment, and transportation costs, palm oil biodiesel is among the least sustainable biofuels. Given the environmental barriers, the dubious energy balance, and the difficulty in using pure biodiesel as a fuel, it appears that most biodiesel will continue to be used as an additive to conventional diesel fuel, at least in the short term.

Biofuel from Waste

Energy sourced from waste products has long been utilized for power generation, often in the form of trash burning (Table 10.2). While thermal use of waste is well known, other forms of waste usage are possible, including municipal solid waste (which includes most municipal garbage), industrial byproducts, and agricultural waste. Common industrial byproducts include wood pulp and sawdust from the lumber and paper industries, already widely used in thermal electricity generation in areas near their production. Agricultural waste products, such as corn husks, sugarcane bagasse (the woody and fibrous parts of the plant that cannot be processed), and the stems or root systems of many other agricultural products are sometimes also burned for power in thermal plants, but only a small percentage of the possible amounts of these is used.

Thermal generation is the primary method of WtE use worldwide. Incineration of municipal solid waste has been an important method of electrical generation for many decades, especially in Europe (Image 10.11). Much as with fossil fuel generation, an incinerator burns municipal solid waste to heat water and drive a turbine, generating electricity. Older incinerator types had efficiencies similar to coal plants (up to

Table 10.2 Generation of Electricity from Biofuels and Waste in the United States

Generation of electricity from biofuels and waste (in gigawatt-hours, Energy Information Administration)

1960	87
1971	258
1973	297
1980	457
1990	86,362
2000	71,713
2010	75,429
2011	77,492

Note: The share of electrical generation in 2011 from biofuel and waste was under 2 percent of total energy usage in the United States.

Image 10.11 Waste-to-energy plants burn trash and waste organic products to run a conventional thermal power plant.

Bernd Neeser / Shutterstock

30 percent), but cogeneration systems to provide heat as well as power can raise this to closer to 80 percent (although the same process can be used in any other thermal generation system). In Europe, strict environmental emissions standards have pushed further developments in clean municipal solid waste incinerators, meaning that ash, particulates, and dioxins are largely removed from emissions, making them a relatively clean method of generation. Despite the improved efficiency of WtE plants, they still attract criticism from many areas. By burning trash that could otherwise be recycled, incinerators destroy a resource that could potentially be valuable. Waste-to-energy systems save land from being used as landfills, but the tradeoff among this land savings, emissions, and reduction in recycling is difficult to quantify. At the very least, WtE systems have not reached a scale large enough to overtake fossil fuel or renewable energy generation, so they remain primarily a method of reducing waste and augmenting the electricity supply.

Similarly, thermal use of bagasse and other agricultural waste products can augment an electricity supply through WtE, although other uses for agricultural waste products may prove more useful and sustainable in the long run. New technologies to use anaerobic digestion or fermentation to convert agricultural waste into ethanol are in development, and they offer several advantages over simply burning the waste. These processes directly convert waste into a usable transportation fuel, which may be more environmentally friendly than burning it locally. Further, the use of agricultural waste to create ethanol, rather than taking food crops to do so, will not impact food prices or supply because the source of the fuel would have been discarded or burned anyway. This effectively means that what was once waste can be processed into ethanol, avoiding many of the problems with a large-scale ethanol economy discussed further in this chapter.

VIGNETTE
Algae as an Energy Resource

A key problem with many biofuels is their intensive use of farmland. Land used to produce agriculturally derived fuel can no longer be used in the production of food and additionally requires large inputs of water and fertilizer. As a fuel crop, algae solves many of these problems: tanks to grow algae at an industrial scale could theoretically be located anywhere, saltwater can be used to grow algae, thus saving fresh water for municipal or agricultural use, and the energy density of algae means it would require less overall land to be used for growing operations. Municipal wastewater could also be potentially used for this production.

A key challenge of terrestrial crop fuels is achieving an energy balance between input and output; for example, in producing corn ethanol, a great deal of water, fertilizer, and external energy input (driving tractors, operating machinery) is required. For many crops, the amount of energy invested in production can approach the amount of combustible energy gained at the end of the cycle, making some fuel crops inefficient from a total energy standpoint. Algae offers an attractive alternative, with its greater energy density than land-based crops (Image 10.12). To convert a plant fuel into biodiesel, ethanol, or other fuels, an important component is the energy content of a given crop, stored either as lipids in the case of biodiesel or as sugars and starches in the case of ethanol. These energy-rich molecules are the primary component that is turned into fuel, while other parts of the plant such as stems, fibrous tissue, or protective husks are produced as waste. Most land-grown crops register under 10 percent lipid content, while certain strains of algae have over 70 percent. Consequently, for a given energy input, algae is already many times more efficient at producing fuel precursors. In addition to lipid content, in ideal circumstances, algae grows much more quickly than land-based crops. Compared to the highest-yield land crops, certain algae types produce almost twenty-five times as much fuel per hectare per year. Oil palm, one of the more efficient terrestrial crops, can produce over 5,000 liters per hectare per year, while high yield algae can produce over 100,000 liters per hectare per year.

While the theoretical yields of algae are clearly higher than that of land-based crops, there are many obstacles to making it a viable energy source. Within the several methods of farming algae, initial costs are generally

Image 10.12 Algae can be grown in special bioreactors, which may be used if a way to harness it as a biofuel is developed.

Toa55 / Shutterstock

high. The most common current method uses a **photobioreactor**, which is a sealed system of glass or clear material that protects the algae from contamination while allowing CO_2 and other nutrients to be easily pumped in. These reactors produce decent energy density per area of land (compared to conventional agriculture) and protect the algae, but the initial construction cost is far higher than that of conventional agriculture. Open growing ponds have been investigated for use as well and offer better cost advantages, but do not protect algae from contamination and have a much lower growth rate per area. Once grown, algae must then be processed into fuel through multiple refining processes. A small number of companies have been undertaking algal fuel production at a research scale with intent to grow, although none of these ventures has taken off.

Assuming that the technical challenges can be overcome, algae cultivation offers a possible advantage over terrestrial crops in both energy input and land use. To grow efficiently, algae requires CO_2 and nitrogen to be pumped into the growth medium, which requires a great deal of energy. Current algae producers have begun using nearby industrial plants as a CO_2 source, although the harvested CO_2 is released into the atmosphere after the biofuel is combusted. Nitrogen is easily harvested from the atmosphere, making it simple to obtain. Additionally, because algae cultivation can use saltwater for growth, plants can be built in places where other crops cannot be grown, such as deserts or coastal areas with no agricultural presence. In this way, algae would not compete for land or resources with food growth, alleviating one of the main concerns with ethanol and other fuel crops. As interest in ecologically sustainable fuel crops increases and the commerciality of algae operations is more concretely established, algae may become a viable transportation fuel. A key concern for a fuel source like algae will be its ability to compete with other developing technologies, in particular solar electric generation. If electric cars begin to take up significant market share and panels continue to drop in cost, land that might otherwise be used for algal fuel production might much more economically be dedicated to solar electric generation.

Worldwide Biofuel Production

In terms of electric generation, biomass (including waste to energy) comprises around 35 GW of installed capacity worldwide. Including biofuels for transportation such as ethanol and biodiesel, Brazil and the United States lead in production by a large margin as a result of long-standing government subsidies and institutional support of biofuel industries. Laws mandating greener transport fuels and CO_2 emissions reductions have driven the expansion of biofuel production in Europe as well, supporting major growth in the industry. World production growth slowed during the economic recession, and some subsidies have lapsed in the United States, but overall biofuel production and usage are likely to at least stay relatively constant in the coming years.

Apart from transportation fuels such as ethanol and biodiesel, many other countries produce significant amounts of biofuels for electricity or heating, but these volumes are much more difficult to estimate with precision, especially global consumption of wood for heating in developing regions. Production of palm oil has been a leading cause of deforestation in countries such as Indonesia, while much of the deforestation in Africa has been driven by wood harvesting for biomass. The use of these natural resources as biofuel is significant, both environmentally and economically, but not quantifiable in a way that allows for cross-regional comparisons.

Biofuel Production in the United States

The United States has become the largest biofuel producer in the world in the past decade, primarily because of a major boom in corn-based ethanol production (Table 10.3). The United States also manufactures ethanol from other agricultural products such as soy and has begun to produce biodiesel in significant quantities as well. The U.S. market for biofuels saw its major boost from rapidly increasing oil prices in the early 2000s, leading to public calls for domestically sourced fuel. Government subsidies for the large corn industry made corn an attractive source for ethanol refining, and major political support for the construction of ethanol plants in corn-producing states led to rapid increases in ethanol production. The United States has become the largest producer of ethanol in the world, now producing more than twice as much in total as Brazil, the long-time leader in ethanol production and use. As of 2011, the United States produced 51 percent of the world's biofuel supply, a share that has increased since that time.

Ethanol has become a significant fuel in the United States, mainly as an additive to standard gasoline, usually up to 10 percent of the total volume. The addition of ethanol to gasoline can help reduce emissions and increase the reliability of supply when there are oil market fluctuations. With the large increases in ethanol production and supply, the use of majority-ethanol fuel (generally E85, or 85 percent ethanol) has also increased. American car manufacturers have responded to the supply of ethanol with dual-use (so-called flex fuel) engines that can run on standard gasoline, E85, or a mixture of both. Particularly in the agricultural regions of the Midwest, where most corn ethanol is made, E85 has become common. While the total use of ethanol is still under 10 percent of the total automobile fuel in the United States, it has been growing steadily, particularly as a blended component of conventional gasoline to reduce urban air pollution.

Table 10.3 Ethanol Production in the United States, 2007–2011 (thousand barrels per day)

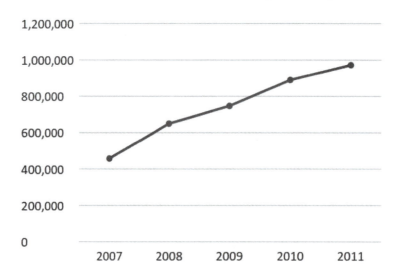

Environmental Benefits and Impacts

Because biofuel includes such a wide variety of sources, the range of environmental benefits and impacts is also wide. Additionally, government support and economic conditions further complicate an assessment of environmental impacts because some sources (such as corn ethanol) are preferred for their economic benefit to farmers more than a strict environmental evaluation would suggest. Broadly, land use changes and emissions reduction are the two main categories for evaluating biofuels. Agricultural biofuels are particularly significant in how they affect land use and agricultural resources, while biomass and waste energy may offer economic benefits although they remain problematic in terms of emissions of carbon and other compounds.

Land use for energy crops has become a major point of environmental controversy around the world. Deforestation of rainforests and jungles in countries such as Brazil and Indonesia for the production of oil palms has led to a massive shrinkage in worldwide forest cover (Image 10.13). This cover is a major natural carbon sink and further protects against erosion, water pollution, and acid rain. When biofuel production is encouraged in countries with dense forest cover, the environmental benefit of producing natural fuel must be weighed against the desire to protect nature and the impacts of deforestation. Because of the fragility of many tropical ecosystems, this tradeoff may prove disastrous. When land use changes are factored in for Brazilian ethanol production, a 60 percent reduction in GHG emissions over oil combustion is achieved, making them generally environmentally friendly in the long run. Further reductions in forest cover, as well as the impacts on biodiversity, make a continued growth of ethanol use in Brazil a difficult environmental problem to calculate even if GHG emissions are reduced in the

Image 10.13 Deforestation in tropical lands to grow sugarcane is a major concern with biofuels. Other land use changes for growing biofuel crops can also be devastating to the environment.
Guentermanaus / Shutterstock

process (Image 10.14). As with many other energy issues, the tension among conservation, emissions, and economics must be resolved according to local factors. For the moment, it appears the Brazilian ethanol production will continue to grow.

In developed countries such as the United States, biofuel production is primarily controversial for two reasons: its impact on global food prices and its energy efficiency. The impact of food crops such as corn being diverted to produce ethanol has possibly increased corn prices worldwide, which has disproportionately affected less developed countries. While this is generally not an economic problem for the United States itself, the moral justification of depriving food security for others in exchange for transportation fuel is challenging. Another separate but equally controversial issue of ethanol use regards its efficiency as a fuel. Corn is the largest recipient of agricultural subsidies in the United States, with nearly $10 billion in annual payments. The large subsidies given to corn growers distort the market, making it a more attractive crop to turn into fuel than it would be without government help. Direct ethanol subsidies expired at the end of 2011, yet insurance subsidies and direct payments for corn continue.

For WtE biofuels and biomass, the primary environmental impact is that of emissions because combustion releases GHG and particulates. The CO_2 from certain types of biomass is recently fixed, so its impact is carbon neutral (although biomass systems like algae use carbon from fossil fuels and are thus not carbon neutral). Without advanced filtration and scrubbing systems, however, combustion-based energy still releases particulate pollution and is therefore not completely clean. Biomass for industrial-scale energy is not widespread, so its total impact on global emissions is low. Further, the most intensive use of WtE systems is in Europe, where advanced technologies have mitigated the majority of harmful pollutants. Because its global use is so small compared to that of fossil fuels, the overall impact of biofuel for energy generation is limited.

Image 10.14 Brazil leads the world in sugar ethanol production.
Alf Ribeiro / Shutterstock

VIGNETTE

Food or Fuel?

A common objection to biofuel production is that food crops can be diverted for use as energy crops, which may have negative effects on food prices and availability for poorer people who might depend on the crops as a staple. This issue became a major focus in 2008, when world food prices rose sharply, leading to food scarcity in many places across the world (Image 10.15). Many observers alleged that biofuel production from corn and soy in the United States was a major cause of the rise in food prices, although later analyses have shown that this impact may have been overstated.

Between 2007 and 2008, many factors such as high oil prices, droughts, higher demand for exported food, and economic speculation contributed to massive price spikes of many different food staples across the world. Wheat, corn, soy, and rice prices rose by double or more in the span of less than a year, leading to major food scarcity in many developing countries (Image 10.16). Riots and political instability followed, and many felt that the increasing use of food crops as biofuels in the United States and Europe was contributing to the price rises. The idea that biofuel production was a major component of food scarcity subsequently led to calls to reduce or stop biofuel production from food crops altogether, with some suggesting that only crops like sugarcane in Brazil or the still-developing technologies for plants like switchgrass should be used for biofuels.

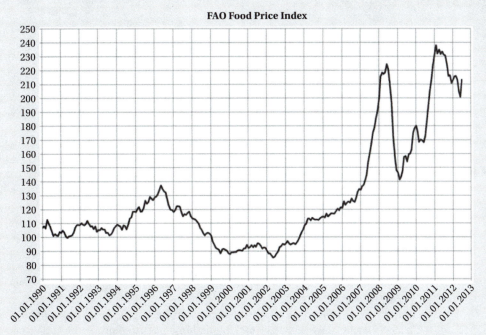

Image 10.15 Food and Agriculture Organization Food Price Index

Source: Food Price Index, 1990–2013, United Nations Food and Agriculture Organization. Available at http://www.fao.org/fileadmin/templates/worldfood/Reports_and_docs/Food_price_indices_data.xls

continued

continued

Image 10.16 Corn is used to produce many food products. When redirected for fuel production, food prices for corn-based foods increase.
Ildi Papp / Shutterstock

Although many factors contributed to the 2008 food crisis (which resurfaced again in 2011), biofuel production certainly played a role, even if it is unknown how much it or other factors contributed to the crisis. This kicked off the debate of *food or fuel*—namely, the idea that food crops should not necessarily be used as direct energy sources, especially when it may deprive others of food security. This debate raises larger questions about energy, sustainability, and our increasingly globalized world. If energy use in a developed country leads to starvation or deprivation elsewhere, is it worth the potential gains in sustainability or emissions reductions?

REVIEW QUESTIONS

1. Why is it so difficult to quantify global biofuel consumption compared to other energy resources?
2. Proponents of biofuels promote them as "green" energy sources like wind or solar. What are some ways in which biofuels are not so green?
3. What are some of the major drawbacks of WtE electrical generation? Is it carbon neutral?
4. What role does politics play in the adoption of ethanol and biodiesel as a transportation fuel option? Why is it so concentrated in two countries?
5. Why is sugarcane a far better feedstock for ethanol than corn? How would the introduction of large-scale cellulosic ethanol change this calculation?

Image 11.1
Poul Riishede / Shutterstock

CHAPTER 11

Geothermal Power

Heat from deep within the earth's interior holds great promise for limitless heat and electricity where geological circumstances allow. Since ancient times, people have used hot springs and other geothermal features for bathing, cooking, agriculture, and heating. To this day, hot springs attract people to their naturally occurring warm waters, which some believe to have strong healing and medicinal properties (Image 11.2). Today, we can harness **geothermal energy** for generating electricity and have developed new ways to heat and cool buildings using natural temperature gradients in the earth's crust. Electricity was first generated geothermally in 1904 and in recent decades has grown significantly as an energy resource.

Depending on the geology in an area, permeability of the rock, and how much water is in the rock, heat flows from deep underground toward relatively shallow areas. In a favorable location, for example, where the recharge rate is fast enough, we can use this heat for our purposes. Overuse can limit future output, but only for a short period until the heat is recharged.

This chapter will examine the physical processes that produce heat deep within the earth, explore the different ways in which society harnesses this energy for heating and electrical generation, investigate the environmental impacts of geothermal energy, and explain geothermal energy's share of the energy basket of the United States and the world.

Image 11.2 Hot springs are found in areas with higher geothermal gradients. People have used the warm waters in many ways over history.
Iamnong / Shutterstock

Geothermal Overview

While most renewable and fossil energy resources can be tied to solar energy, geothermal—along with nuclear—derives its energy from the internal heat of the earth. Geothermal energy is virtually inexhaustible, relatively sustainable, and generally has a constant output. In the past, geothermal exploitation has been geographically limited. However, enhanced geothermal extraction may allow geothermal production in many and perhaps most places.

The physical processes of the earth's interior are the key to understanding how geothermal energy is produced and exploited (Image 11.3). At the most basic level, heat from over 2,800 kilometers below the surface, deep within the earth's core, flows outward. Because of the latent heat from the formation of the planet, as well as radiation, the **inner core** measures approximately 5,430°C. This intense heat flows through the **outer core** and into the **mantle** and **crust**. About 10 percent of this heat dates back to the formation of the earth, while the other 90 percent comes from the decay of radioactive elements in the core, including uranium, potassium, and thorium. The mantle is the thickest of the earth's layers, consisting of a semisolid state that is viscous enough to allow heat to flow. At depths of 80 to 100 kilometers, the mantle's temperature is still about 650°C to 1,200°C, while in the crust at depths of around 10 kilometers, the temperature averages only 300°C.

Despite the heat within the mantle being distributed throughout, the ability to harness this heat at the surface varies widely. **Oceanic crust** is thin, as little as 5 kilometers thick, while **continental crust** is much thicker, ranging mostly from 30 to 50 kilometers thick. Because geothermal projects so far have been located on land, the thickness of the crust in most places impedes our ability to harness the massive heat of the mantle

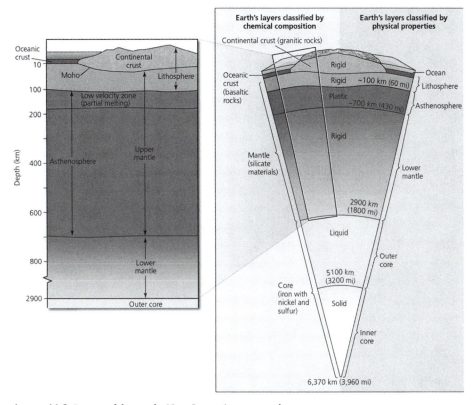

Image 11.3 Layers of the earth. *Note*: Image is not to scale.

in amounts sufficient for large-scale power generation. In the crust, the typical **geo-thermal gradient** is about 3.3°C per 100 meters, so in most places if you dig down 100 meters, the temperature you find will be 3.3°C hotter than what you find at or near the surface. However, this gradient varies widely from place to place. Some places have much higher gradients because of specific geographic factors, such as in Iceland, where gradients can reach as high as 30°C per 100 meters.

Certain geological conditions cause much higher thermal gradients and are there-fore the primary areas in which geothermal electricity can be economically gener-ated. Areas of especially active volcanism are the most apparent places, especially along tectonic plate boundaries, where cracks between plates cause a much higher incidence of volcanism and geothermal activity, such as the **Pacific Ring of Fire** (Image 11.4). So-called **hot spots**, such as Iceland and Hawaii, are also excellent geo-thermal energy locations. Hot spots are likely caused by massive magma intrusions direct from the mantle, but many lie on the ocean floor and are difficult to harvest for geothermal energy.

Places where there are many hot springs, geysers, and volcanic features are often called a **geothermal reservoir** (Image 11.5). In such places, there is a strong thermal gradient, where areas close enough to the earth's surface to be harnessed are heated

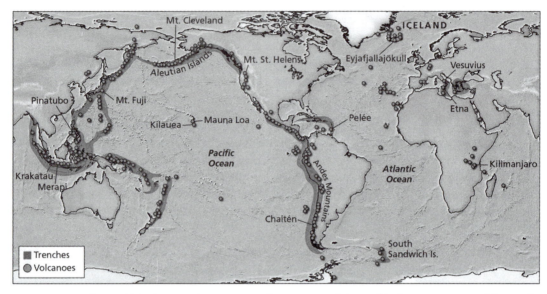

Image 11.4 This map shows recent volcanoes around the world. The majority around the Pacific Ocean form the Ring of Fire.

Image 11.5 Yellowstone National Park is home to the largest collection of geothermal features anywhere on the planet.

to temperatures above the boiling point. In these areas, groundwater seeps down to the superheated rocks below, where it is heated and comes back to the surface. The geothermal gradient is a primary concern for locating large-scale projects. Geothermal sources are usually classified into four temperature categories: high (greater than 160°C), medium (90°C–160°C), low (30°C–90°C), and very low (less than 30°C). Electricity projects are only appropriate in high- and medium-grade resources, while heat for domestic hot water and building heating can be produced in all four resource types.

Just as important for conventional geothermal production as the temperature of the rock is the permeability of the rock formations. Permeable rock formations allow water to move freely within them, so for water to be cycled between the heated rock and the surface, high permeability is essential.

Water supplies are also important in geothermal plants. While most permeable rock reservoirs will initially contain water, a geothermal project can deplete the water over time. In fact, in 1999 at the Geysers geothermal plant, production had fallen significantly simply because the water used to move the heat out of the rocks and up to where it is usable was exhausted. Geothermal output was brought back by injecting new water into the geologic formation—water from a nearby sewage treatment plant. In fact, most modern geothermal generation includes reinjection of water to maintain output.

Beyond geothermal power projects and large-scale heating projects in especially active geothermal regions, geothermal energy can also be used to heat and cool buildings. Rather than taking advantage of the high temperature of subsurface reservoirs, the stable temperature of the shallow crust can be used. On the ground surface, solar heating leads to large fluctuations in temperature from day to night and between seasons. However, just under the earth's surface, temperatures remain constant in most locations, allowing for building heating and cooling systems that take advantage of that just-underground constant temperature.

Geothermal Technologies

The three principal applications for geothermal energy are as follows: (1) direct use and heating, (2) electrical generation, and (3) **ground source heat pumps**. Direct-use systems utilize naturally occurring hot springs on the surface or subsurface of hot water reservoirs for the purposes of heating water or manmade structures. Geothermal power plants make use of a geothermal reservoir to heat water into steam, which can then be used to run thermal generators, as in other conventional steam-driven electrical plants. Ground source heat pumps take advantage of stable ground temperatures for climate control in buildings. Heat pumps can use low- or very-low-grade geothermal resources, while direct-heat systems require medium- or high-grade resources. Each of these systems is described in further detail below.

Direct-use systems are the simplest of all geothermal technologies. For water heating, piping water directly from naturally occurring hot springs or subsurface reservoirs is all that is required. In Iceland, where there is an enormous hot spot underlying most of the country, direct-heat systems also use piped hot water to heat buildings during the winter. In Reykjavik, Iceland's capital, a district heating system provides heating to about 95 percent of all buildings in the city and even is used to melt ice and snow on sidewalks.

Geothermal power plants come in a variety of forms, but all must use a geothermal reservoir of sufficiently high temperature. In general, these plants require a heat source of between 100ºC and 370ºC. Wells that pipe hot water or steam to the surface are drilled from about 1.5 to over 3 kilometers deep. Because these underground sources are always hot, geothermal electricity is always available. This is particularly useful in a grid powered with variable wind and solar electricity.

The simplest type of geothermal power plant is the **dry steam plant**. These plants take advantage of naturally occurring steam, which is tapped and piped from an underground source to the surface and then run through a conventional thermal turbine. The world's first geothermal power plant in Larderello, Italy, was of this design, as is the largest U.S. plant, the Geysers (Image 11.6).

The second type of geothermal power plant, and the most common, is the **flash steam plant**. These plants take hot water from underground that is piped to the surface as steam to run a turbine. The steam is then cooled and condensed back into water and piped back underground and recycled through the plant. In certain places where there is little groundwater, the process is entirely fed by the injection of surface water.

The third type of geothermal plant is the **binary geothermal power plant**. These plants use a heat exchanger, where another liquid with a low boiling point is run in pipes adjacent to the geothermally heated water. This other liquid is then boiled into a gas and used to drive the turbine and then condensed and reused.

Image 11.6 The Geysers is the largest geothermal power plant in the United States. Courtesy of Ted J. Clutter / Science Source.

Ted J. Clutter / Science Source

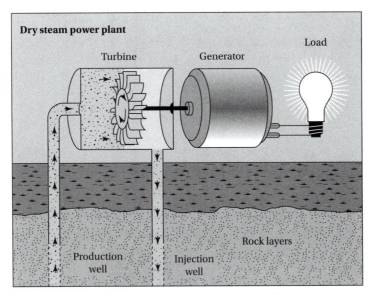

Image 11.7 This chart shows a dry steam geothermal power plant.
Courtesy of the U.S. Energy Information Administration

A relatively new use of the temperature gradient available just below our feet is the ground source heat pump, used to heat and cool buildings and heat water (Image 11.8). In recent years, there has been a widespread movement to install heat pumps in more buildings. Rather than using the high heat of deeper geothermal reservoirs, a ground source heat pump uses the thermal gradient between ambient air and the shallow ground (as little as 3 meters deep) as a heat exchange for climate control. At this depth, most areas of the ground in the midlatitudes stay constant at around 10°C to 16°C. This means that in summer, the ground is cooler than typical outside air temperatures, while in winter it is warmer. A heat pump takes advantage of this difference to transfer heat from the ground to buildings in winter and to transfer heat from buildings into the ground in summer.

Ground source heat pumps do not take advantage of the earth's heat the way their deeper drilling cousins do. The ground temperature in most places only a few meters deep is more established by the average of a years' air temperature than the temperatures found by drilling deeper. Rather than being a pure renewable, air and ground heat pumps act as an energy amplifier. They use a small amount of electricity to move around a lot of heat. A typical air source heat pump moves two to four times as much heat energy as the electric energy used (e.g., 1 kWh of electricity leads to the production of 2–4 kWh of heating or cooling). A typical ground source heat pump moves three to six times as much heat as the electricity used. Studies by the U.S. Environmental Protection Agency have shown that heat pumps are among the most energy-efficient, clean, and cost-effective systems for building climate control, although they cannot be used in all circumstances.

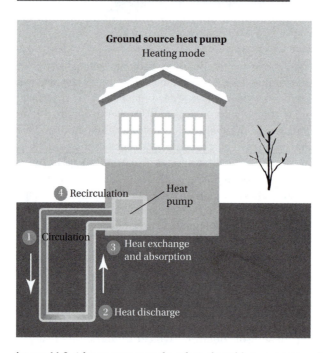

Image 11.8 A heat pump uses the relatively stable temperatures found underground as a heat exchanger to help cool buildings.

Courtesy of the U.S. Environmental Protection Agency

Environmental Impacts

Compared with most other energy resources, the impacts from geothermal power are low. When hot spring water is piped out for use in water or direct heating, there is almost no discernible impact. However, some consequences are associated with geothermal power plants.

Geothermal power plants are not completely carbon neutral. They release a small amount of carbon dioxide into the atmosphere, albeit at a rate that is less than 1 percent of that released from any fossil fuel–powered facility. Additionally, geothermal plants require scrubber systems to remove sulfur dioxide, which is a common byproduct in areas of volcanic activity. When properly mitigated, 97 percent less sulfur dioxide is released compared to conventional coal plants that generate the same amount of electricity. By reinjecting water and steam, other potentially harmful compounds are kept in isolation, rather than being released into the atmosphere.

Placement remains a key concern for geothermal power plants. Systems that utilize the capture of steam and/or reinjection can disrupt naturally occurring hydrothermal features. In New Zealand, geothermal development has already destroyed most of the features in the Wairakei and Tauhara geyser fields. In the United States, strong federal protections over certain features, such as those contained in Yellowstone National Park, have led to the denial of permits for geothermal facilities in nearby areas (Image 11.9). Construction of a geothermal power plant could disrupt or even destroy rare hydrothermal features, such as geysers, located in nearby areas associated with the same geothermal reservoir. In the case of Yellowstone, the area is one of the best suited anywhere in the world for geothermal power production, but is also home to over half the world's geysers and numerous other

Image 11.9 The construction of a geothermal plant can lead to geothermal features, such as geysers, being destroyed nearby.

Lorcel / Shutterstock

hydrothermal features of world renown. As such, there will be no development anywhere close to Yellowstone. Similar sites in Iceland, Northern California, and New Zealand could face similar destruction if geothermal plants were to be built in the area.

Geothermal Use Worldwide

Currently, geothermal power plants operate in twenty-four countries around the world, with a total installed capacity of about 11.8 GW. The United States is the number one producer in total output (although the percentage of total energy generation is small), followed by the Philippines (Table 11.1). The Philippines generate 27 percent of their electricity from geothermal, while Iceland (the world's seventh biggest producer) generates 30 percent of its electricity from geothermal, the highest rate in the world. Indonesia has the highest potential for geothermal power generation because of the amount of volcanic activity throughout its islands.

Geothermal makes up only a fraction of total global electrical generation, and its future is only bright in areas with a specific geography. Much of this potential lies around the margins of the Pacific Ocean, known as the Ring of Fire. Major generators including the United States, the Philippines, New Zealand, and others take advantage of this geography already, while many other countries with major potential, such as Chile, do not generate significant amounts of geothermal power yet. Unlike many other alternative energy sources, many of the major generators of geothermal electricity are developing countries, including three of the top five global generators. Because of its consistency, geothermal can provide an excellent source for electricity in volcanically active places, thus providing a clean and relatively cheap form of power. In places such as El Salvador, these projects have been funded through international aid programs in an effort to raise the standard of living through the provision of cheap and clean electricity. Chile, Kenya, and Indonesia are widely cited as major prospect

Table 11.1 Top Ten World Geothermal Power Generators (in megawatts installed capacity)

United States	3,567
Philippines	1,930
Indonesia	1,375
Mexico	1,069
New Zealand	973
Italy	944
Iceland	665
Turkey	637
Kenya	607
Japan	533
World total	13,300

Source: Geothermal Energy Association. Available at http://geo-energy.org/reports/2016/2016%20 Annual%20US%20Global%20Geothermal%20Power%20Production.pdf

sites for additional geothermal power generation (Images 11.10 and 11.11). Even if brought to optimal use, geothermal is not expected to be able to rise above about 1 percent of global electricity production.

Most global use of geothermal power is not for electrical generation. Approximately three-quarters of geothermal use is in direct applications, such as home and water heating. The total amount of energy harvested for these uses is much harder to measure and much of it is used in rudimentary ways. Certain highly developed countries such as Iceland have built extensive projects that provide heating from geothermal resources at a grid level.

Image 11.10 Indonesia is home to many volcanoes and has huge potential for geothermal power.
Byelikova Oksana / Shutterstock

Image 11.11 This geothermal plant is located on a volcano in Guatemala, along the Ring of Fire.

Geothermal Use in the United States

In 2012, the United States generated almost 17 billion kWh of electricity from geothermal power plants, making it the world's leading generator. Despite this, geothermal power only made up 0.4 percent of total U.S. electrical generation. The geothermal potential of the United States is mostly concentrated in the western states (including Alaska and Hawaii), where most geothermal reservoirs are located (Image 11.12). California leads U.S. geothermal power generation, with 80 percent of the national total (Image 11.13). This includes the largest geothermal power plant in the United States, the Geysers, located north of San Francisco, with a capacity of 1,967 MW. Nevada sits in second place, with 16 percent of national generation, Utah is in third, Hawaii in fourth, and Oregon in fifth place. Idaho is the only other generating state; however, there are plans for small amounts of future generation in Alaska, Colorado, New Mexico, Arizona, Texas, and North Dakota.

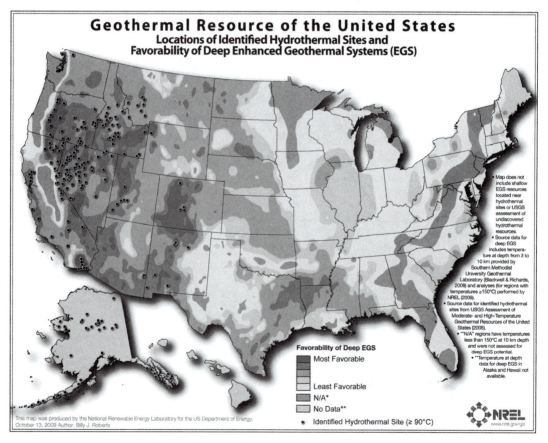

Image 11.12 Only small parts of the United States have strong geothermal potential for constructing power plants.

Courtesy of the National Renewable Energy Laboratory

The U.S. Geological Survey estimates that the United States could easily generate at least three times the country's current production levels. However, because of a lack of investment and other political issues, geothermal electricity is not growing quickly in the United States. Only Hawaii is currently seen as a location for large amounts of additional capacity, where there is a proposal for a 50 MW facility that would displace current generation from oil and diesel power.

Even at three times the current generation level, geothermal will only play a small role in the overall U.S. energy basket for electricity. Consequently, most of geothermal's future promise comes from its other uses. Most new capacity in the United States is in the home cooling and heating realm, where heat pumps are becoming more and more popular. As a result of tax credits and rising energy costs, geothermal is becoming a more attractive means for Americans to heat and cool buildings. Many new environmentally certified buildings make use of geothermal climate control systems as one means of gaining this certification.

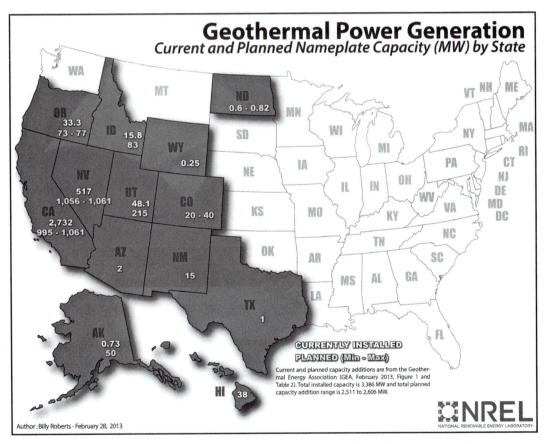

Image 11.13 Almost all geothermal power plants in the United States are located in a small number of western states.

Courtesy of the National Renewable Energy Laboratory

VIGNETTE
Geothermal in Iceland

Iceland offers a unique look at the use of geothermal energy in a geographic location well suited to large-scale geothermal development. Located along the Mid-Atlantic Ridge, Iceland sits on a major hot spot, with twenty-six high-energy geothermal hotspots under the island country (Image 11.14). Access to this resource, along with extensive hydropower resources, allows Iceland to be the world's highest per capita energy consumer, yet also one of the world's greenest.

Geothermal resources constitute over 50 percent of Iceland's total energy basket, making Iceland by far the highest per capita consumer of geothermal energy in the world. For geothermal electricity specifically, Iceland derives 30 percent of its total generation from geothermal (the rest comes from hydroelectric plants) (Image 11.15). The majority of geothermal in Iceland is used in direct applications, providing heating for buildings during Iceland's long and cold winters, as well as hot water heating and even snow and ice melting.

Iceland's history with geothermal dates well into its past. Hydrothermal features were long used in Iceland for purposes of bathing and washing. By the early twentieth century, Icelandic farmers were using hot spring water to heat greenhouses, thereby extending their short growing season. These uses remain today, with about 3 percent of geothermal consumption in both areas, including extensive use for heating swimming pools. Swimming outdoors remains a national pastime in Iceland because of geothermal energy, despite its northerly climate, and one of Iceland's major tourist attractions is the Blue Lagoon hot springs.

By the 1930s, Iceland employed geothermal energy for district heating, first in the capital city of Reykjavik. By the 1970s, this use had spread around the country, and today twenty-six local district heating utilities provide space heating from geothermal sources, which together heat almost 90 percent of Icelandic homes.

Geothermal electricity was first introduced to Iceland in the early 1970s and took off in the mid-1990s. This electric source replaced imported coal and oil for power production, which, along with hydroelectric, has both allowed Iceland to become independent for electrical production and allowed electrical production to be green.

Image 11.14 Iceland is located on the Mid-Atlantic Ridge and has enormous geothermal reservoirs.
Ovchinnikova Irina / Shutterstock

Because of its unique geography and climate, Iceland is a unique case in the realm of geothermal energy potential (Image 11.16). Its abundant geothermal resources allow it to be one of the most sustainably powered countries, while still enjoying a high standard of living with large per capita energy use. Additionally, Iceland uses its abundant geothermal electricity to produce much of the world's aluminum. Refining aluminum from bauxite (the mineral it is generally produced from) is highly energy intensive, so the cheap and clean electricity in Iceland allows it to produce aluminum both more cheaply and more sustainably than in most other places. While most countries could not replicate Iceland's example as a whole, their innovation in geothermal technology has been an important force in the worldwide exploitation of geothermal resources for electricity and energy.

Image 11.15 A geothermal plant in Iceland.

Jose Arcos Aguilar / Shutterstock

Image 11.16 Iceland is famous for its geothermal features. The word geyser comes from the Icelandic language.

Dennis van de Water / Shutterstock

REVIEW QUESTIONS

1. What are the key considerations in determining the location of geothermal projects? Also, list the classifications of geothermal sources and which ones can be used for electricity production.
2. Why is geothermal a more effective base load power generation option compared to wind and solar power?
3. Describe the environmental impacts of the various systems within a geothermal power plant.
4. How does Iceland's unique energy basket influence its position as the world's highest per capita geothermal energy consumer?
5. What are the geographic limitations of geothermal energy production? Where is its future the brightest?
6. If geothermal is brought to its optimal level of use, what percentage of the total global electricity production will it represent, and how is most geothermal power used?
7. Describe the geographic distribution of geothermal energy production in the United States today and for future projects. What challenges does the future of geothermal power development face, and what incentives are given to Americans to adapt new uses of the power source?

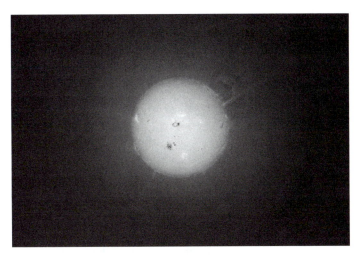

Image 12.1 Courtesy of Detlev van Ravenswaay / Science Source.
Detlev van Ravenswaay / Science Source

CHAPTER 12

Fusion and Other Potential Future Technologies

The previous chapters have laid out the major sources of energy that are consumed to currently meet humanity's energy needs. All have drawbacks and none is a silver bullet that will meet our vast energy needs without serious impacts on our society, politics, and environment. As work continues to improve the means by which we consume these resources, research also is moving forward on future technologies that could either supplement or replace current resources altogether.

Fusion, using the same physical processes that power the sun, has long been assumed to be the ideal energy source if technology could advance sufficiently to access it. With the ability to generate almost limitless power with few impacts, fusion would rapidly displace fossil fuel use in the generation of electricity. So far, achieving sustained fusion reactions like those that power the stars has proven elusive, but research continues. Other potential future alternatives are also being investigated. Many breakthroughs have occurred in hydrogen power, especially in developing hydrogen fuel cells, but challenges remain for producing the needed hydrogen feedstock without contributing to global carbon dioxide emissions. Finally, proposals for future power production exist, such as space-based satellites that would harvest microwave or other radiation spectra from the sun and beam them to the earth for power production.

Today, none of these technologies is more than experimental, and many remain purely theoretical. Even so, the multitude of problems, costs, and impacts that result from our current energy basket push scientists to research along many diverse paths. As with other energy technologies, breakthroughs could come at any time, and some technologies discussed in this chapter could have a massive impact on world energy use if the technical and scientific hurdles can be overcome.

Fusion Power

Nuclear fusion is the same physical process that powers the sun and the other stars (Image 12.2). If a sustained fusion reaction could be produced, controlled, and exploited in a power plant, it would represent the holy grail of energy production because it could theoretically produce enough power to meet all of humanity's energy needs. Overall, fusion would represent the single most important advance in power production yet achieved. Fusion has been pursued for decades, often lauded as a breakthrough that is close at hand, yet for all the research and press coverage, fusion technology seems to be perpetually just out of reach.

Fusion would come with many advantages over current resources. Its fuel sources are virtually limitless compared to global energy needs and its byproducts are relatively harmless gases, unlike the highly radioactive byproducts of nuclear fusion's cousin, nuclear fission.

Whereas nuclear fission requires rare, expensive, and dangerous heavy elements such as uranium and plutonium, fusion's feedstocks are far safer and more common. Fusion instead works with the lightest elements, using isotopes of hydrogen that are combined to produce helium. Just as nuclear fission creates heat and light as large elements break apart, fusion produces massive amounts of heat and light as two hydrogen atoms are fused together to produce one helium atom. Similar to nuclear fission, in which isotopes of uranium and plutonium are required for reactors, only the heavy

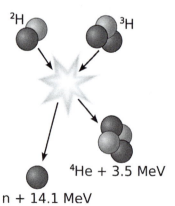

Image 12.2 At temperatures around 100 million degrees Celsius, deuterium and tritium can fuse, creating a helium atom and a free neutron, the latter of which carries a large amount of excess energy (four-fifths of the total from the reaction). Without such high temperatures, the repelling forces of the atoms prevent such high temperatures from being reached.

Courtesy of the Princeton Plasma Physics Laboratory

isotopes of hydrogen are suited for fusion reactions. In this case, **deuterium** (H-2 or hydrogen with one proton, one electron, and one neutron) and **tritium** (H-3 or hydrogen with one proton, one electron, and two neutrons) are both useable. Hydrogen without any neutrons, H-1, is not a suitable feedstock because helium only forms with two protons and either two or three neutrons in its nucleus.

Fusion reactions are created and sustained using different techniques than those that create a fission chain reaction. Fusion does not create a chain reaction at all; rather, it uses extreme heat to cause fusible nuclei to collide and create a **thermonuclear** reaction. When the temperature and pressure are high enough, collisions between hydrogen atoms increase, creating **plasma**, a substance that is both superheated and maintains an overall charge of zero (same amount of electrons and protons when the mass as a whole is considered). Plasma that sustains a fusion reaction reaches temperatures of at least 100 million degrees Celsius (180 million degrees Fahrenheit), causing the molecules to move at speeds one-tenth the speed of light.

Creating what is effectively a miniature star with such extreme conditions and being able to contain it is a significant challenge. One would need to contain the reaction with outside forces that would bottle it up, with some sort of coolant flowing around it that would be used to transfer out heat for the purpose of then generating electricity using conventional thermal turbines. While the theoretical basis for fusion power is sound, three principal obstacles remain: (1) creating and maintaining the heat needed to fuse atoms, (2) containing the plasma in a way that will facilitate continued fusion reactions over time, and (3) maintaining the fusion reactions long enough so that more energy can be extracted than it takes to start and maintain the fusion reaction.

Until recently, the only fusion reactions created by people that have produced net power have come from the hydrogen, or thermonuclear, bomb, which creates an uncontained thermonuclear explosion for a split second using deuterium. Controlled fusion using a thermonuclear reaction of deuterium with tritium is generally favored because it can occur at a lower temperature and density than fusing two deuterium atoms. In stars, fusion reactions are contained because of the enormous gravitational forces of the stars themselves. There are no containers or materials that can hold plasma directly; indirect containment methods are a prerequisite to achieving a sustained fusion reaction on earth.

Plasmas can be contained safely in a magnetic field. A variety of shapes of magnetic fields are being researched, including the torus (a circular or donut-shaped tube). The **tokamak** arrangement uses this torus and magnets cause the plasma to flow with the magnetic field through this tube, which is the model currently favored in fusion research. Two facilities conduct research on this method—the Joint European Torus in the United Kingdom and the Tokamak Fusion Test Reactor at Princeton University (Image 12.3).

The other primary means currently under investigation uses **inertial confinement** to cause an implosion of deuterium and tritium. This is the means used in hydrogen bombs, wherein hydrogen atoms are forced inward on each other. In a bomb, this process is done with fission explosions around a deuterium–tritium core, a task impossible for a power plant. Instead, research is being explored using lasers to fire on a core, primarily at the National Ignition Facility (NIF) at the Lawrence Livermore National Laboratory in

Image 12.3 The JET Tokomak reactor has been built in England
for testing fusion theories. Courtesy of EFDA-JET / Science Source.
EFDA-JET / Science Source

California. The NIF uses the world's largest laser array. In 2014, the NIF achieved a small
net energy gain in a fusion reaction for the first time, an important step in the development of commercial-scale fusion power. This is the only fusion method that has passed
this milestone, and it still remains far below the scale needed for commercial use.

Successful fusion reactions have been created in a number of different ways so far,
yet almost all have required more energy input than they generate. Unless a technology can be developed that is a net power generator rather than a net power consumer,
fusion as a major energy source will remain purely theoretical. Even so, fusion research
continues at many laboratories and companies around the world; if the numerous technical and scientific challenges can be overcome, fusion could become the fuel of the
future. So far, experimental results have failed to prove the viability of the theory, and
only time and further research will demonstrate whether it is possible.

Fuel Cells and Hydrogen Power

Hydrogen, the lightest element and the most common substance in the universe, has
great potential as an energy carrier. Hydrogen is a rich energy source both for direct
combustion and as an ideal method of generating electricity in **fuel cells**. Hydrogen's
lightness is its most critical weakness, however, because hydrogen gas rises quickly in
the atmosphere, with most being ejected into space or quickly combining with other
elements such as oxygen. Consequently, obtaining pure hydrogen on earth requires
the processing of other compounds that contain it, making it a secondary, rather than
primary energy source. As such, energy is required to create pure hydrogen, making it

The National Ignition Facility

Although several labs around the world work on fusion research, none has come further in recent years than the **National Ignition Facility** (NIF) at the Lawrence Livermore National Laboratory in Livermore, California, near San Francisco (Image 12.4). The NIF is the world's largest laser, with over forty thousand optics that focus 192 laser beams onto a target that is about the size of a single pencil eraser.

The NIF program aims to create initial fusion ignition, in which a self-sustaining fusion reaction occurs wherein deuterium and tritium fuse together to form helium, like in a star. Fusion ignition would be the point at which more energy is released in the fusion reaction than was input into the process, that is, more energy out than in. So far, no fusion reaction has yielded this outcome.

The NIF works by taking a single small laser beam, amplifying it by a factor of 10 billion, and splitting it to enter the target chamber as 192 beams. This is done by sending the beam through optics over a course of 1,500 meters (4,921 feet) in an assembly that is the size of three football fields. The laser travels this distance in 5 microseconds! These beams enter the chamber and create X-rays that then compress the fuel target

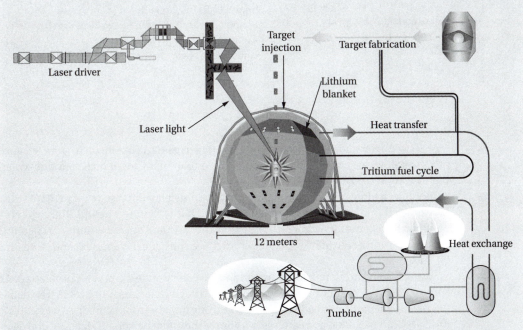

Image 12.4 At the National Ignition Facility, 192 lasers converge on a fuel capsule to produce a fusion reaction. If this can be done efficiently, it could become a prototype for a fusion power plant, as this diagram depicts.
Courtesy of the Lawrence Livermore National Laboratory

continued

continued

or hohlraum, a tiny gold capsule filled with deuterium and tritium, until it collapses in on itself to the point that nuclear fusion takes place (Image 12.5).

In 2014, the NIF accomplished the first fusion reaction in a test where more energy was released than was put into the fuel, an achievement that is one of the most important steps forward in recent fusion research. The overall energy put into the laser system was still far greater than the energy release during the fusion reaction, so this was just one step along the process toward developing fusion power.

The experiments were able to create a density more than twice that of the inside of our sun, which is important to create the plasma necessary for sustained fusion. Although that elusive goal has not been achieved, the work at NIF is yielding results that have taken us closer to fusion than any prior experiments.

Image 12.5 The target chamber at the National Ignition Facility is where the reaction occurs. Courtesy of LLNL/Science Source.

Lawrence Livermore National Laboratory/Science Source

an energy carrier and not a direct source. Hydrogen gas is also difficult to store because of its very low density (it is more energy dense than many fossil fuels, but only when highly compressed).

Because hydrogen does not occur naturally in its elemental state on earth, it must be extracted from chemical compounds in which it is found in nature. The two most common feedstocks for hydrogen production are water and methane. Through chemical processes, hydrogen can be separated from these substances, captured, and stored for later use.

Water is split into elemental hydrogen and oxygen through the process of **electrolysis** (Image 12.6). By passing an electrical current through water, the water is separated into its constituent parts: two hydrogen atoms and one oxygen atom. Chemicals such as sulfuric acid can be added as a catalyst to improve the efficiency of the reaction, and electrodes of rare metals, such as platinum, also aid in the process. No waste products are produced through electrolysis because chemical catalysts can largely be recycled. The primary challenges are the source and cost of the electricity because electrolysis is an inefficient and highly energy-intensive process. If the electricity for hydrolysis can be sourced from renewables such as wind or solar, the process will effectively be emissions free. If, however, the electricity were provided by fossil fuel generation (as a majority of all electricity currently is), the environmental benefit of hydrogen becomes much less

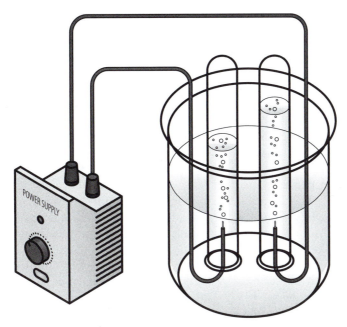

Image 12.6 Electricity is used in an electrolysis reaction to separate water into hydrogen and oxygen gases.

Zern Liew / Shutterstock

compelling. Production of hydrogen gas through electrolysis also results in a net loss of energy because it takes more power to split water than is gained in the end. Consequently, highly abundant and clean electricity is a prerequisite for any large-scale implementation of electrolyzed hydrogen.

Hydrogen can also be isolated from methane gas via the process of **steam reforming**. At a temperature of greater than 700°C (1,292°F), steam and methane are combined in the presence of a nickel catalyst to form hydrogen and carbon monoxide gas. A second reaction at a lower temperature combines the carbon monoxide with more steam to produce more hydrogen, along with carbon dioxide gas. This is the most common method used today for hydrogen production because it is far more efficient than electrolysis. However, this process has major environmental impacts because it requires significant heat input for the first part of the reaction, often accomplished with the burning of natural gas, and the final reaction produces climate-altering carbon dioxide gas, thereby making this form of energy problematic in the same fashion as the classic fossil fuels used for so much energy production today.

In the future, other ways to produce hydrogen in an environmentally friendly manner may be found. Current research includes investigation into natural forms of hydrogen production, such as from algae or bacteria that produce hydrogen naturally, which could be used as an energy resource if captured.

Because most hydrogen supplies go to industrial uses such as ammonia production rather than for energy uses, it is difficult to accurately estimate global production levels. In the United States, about 9 million metric tons of hydrogen are produced per year. According to the Energy Information Administration, this is enough to power twenty to thirty million cars or five to eight million homes. However, the majority of this hydrogen is directed to industrial applications and is not a significant part of America's energy basket at this time.

The one major global outlier using hydrogen as fuel is the space program, both in the United States and in other countries. Liquefied hydrogen has been used by NASA as one of the fuels for its rockets for several decades, and the manned spaceflight programs use hydrogen fuel cells for power while in orbit and then are able to use the water byproduct as drinking water for the astronauts (*Note*: The International Space Station is an exception to this because it uses solar power) (Image 12.7).

The fuel cell is by far the most common way to use hydrogen as an energy source, both for direct electric generation and for transportation purposes. Hydrogen fuel cells work in a similar manner to batteries, but require a constant outside supply of reactive chemicals (hydrogen and oxygen in this case) rather than generating electricity entirely from chemicals stored within it (Image 12.8). Each fuel cell contains an anode, a cathode, and an electrolyte. In the anode, the fuel is converted into an electron and an ion, after which the electrons are passed through wire as electric current, while the ion passes to the cathode and reacts with an electron to form water. In this type of system, electricity and water are the only products, making it clean. If the hydrogen fuel can be sustainably produced, fuel cells offer low-impact portable electric generation for vehicles or building uses. Most fuel cells are currently used to generate building electricity in remote locations and even aboard spacecraft. Because of their mechanical simplicity, these systems are efficient. Increasingly, hydrogen-powered cars have been discussed as a transportation alternative, although development has been slow and most focus is currently on battery-powered cars. Apart from some technology demonstrations and a limited network of a few hundred hydrogen vehicles and filling stations in California, high costs and limited infrastructure have prevented hydrogen fuel cell vehicles from significantly penetrating the market.

Image 12.7 The space shuttle used liquified hydrogen to fuel the launch.

Everett Historical / Shutterstock

Hydrogen fuel cell

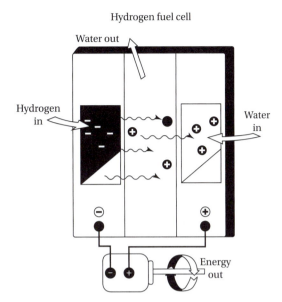

Image 12.8 Hydrogen fuel cells combine hydrogen and oxygen to produce electricity and water.

Courtesy of the U.S. Energy Information Administration

Fuel cells remain expensive, which has been a major barrier to their widespread adoption. Prices will continue to drop as production increases, although some industry analysts suggest that automotive fuel cells could take decades to become cost-competitive with internal combustion engines. Beyond price, infrastructure and transport are perhaps the largest barriers to the widespread adoption of hydrogen as a fuel. Only a few hydrogen filling stations exist, so hydrogen-powered vehicles cannot be used outside these locations. For them to meaningfully compete with electric or gasoline vehicles, thousands of new filling stations and pipelines would need to be built, which would take massive investments that are unlikely, given the uncertainty of the adoption of hydrogen as a fuel. These challenges are not insurmountable, and interest in hydrogen vehicles has spurred many automotive manufacturers to develop new vehicles using the technology, although much of this effort is to achieve government credit for building zero-emission models. The big advantage of fuel cell hydrogen vehicles is that they are refueled much like today's gasoline vehicles. Given the major barriers to hydrogen fuel, however, they likely will not become a major part of our energy basket in the near future.

Other Proposed Energy Sources

Beyond the conventional and alternative sources described in the preceding chapters, many other alternative possibilities have been suggested by researchers over the years. These so-called new alternatives always seem to be a moving target. In the

1990s, proposals were made for microwave-beamed power, in which huge satellites would use solar arrays to capture solar energy outside the atmosphere and beam it wirelessly to earth. Solar energy is more concentrated and constant in space, so if wireless power technology can be developed, this method would allow for continuous and completely clean solar power to be delivered into the grid from space. New ideas continually are proposed, some less likely than others, but the possibility remains that the great solution to our energy challenges may lie with a resource other than those currently produced.

REVIEW QUESTIONS

1. What are some of the barriers to the development of fusion power?
2. How is current hydrogen production a net contributor to greenhouse gas emissions?
3. Could hydrogen conversion be used as a means to store excess power produced by solar and wind power? How would we likely consume this hydrogen?

Image 13.1
Ortodox / Shutterstock

CHAPTER 13

Transportation

In the past two centuries, the rapid development of new engines, primarily powered by fossil fuels, has fundamentally transformed the face of our world. Military power, trade, globalization, and communications have massively increased in scope, and transportation has been at the heart of economic growth for all these areas. In the nineteenth century, the development of steam-powered trains and ships led to faster transportation both on land and at sea. In the twentieth century, the development of the automobile and airplane shrank the world we live in even more. These developments allowed the growth of car-centric megacities and cut intercontinental travel to a matter of hours, fundamentally reshaping the speed with which people could interact with others globally (Image 13.2).

Our transport systems make modern life possible and serve as the backbone of the global economy, but they also have an insatiable appetite for energy (Image 13.3). The energy most methods of transit use must be portable (with the exception of electric trains and certain buses). So far, that has mostly meant fossil fuels. Fossil fuels have served as the backbone of this transportation revolution, starting with coal-fired steamships and locomotives and later the development of petroleum-fueled cars, trucks, trains, ships, and aircraft. Today, these fossil-fueled forms of transportation not only move passengers around the globe, but also move the food we eat (and the machines on farms that grow it) and the various consumer goods we purchase. In fact, cargo consumes most of the energy in global shipping and a significant portion of the energy used in aviation.

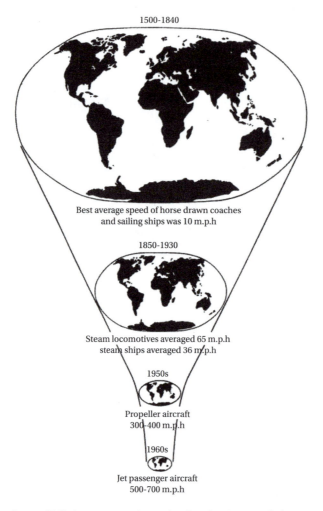

1500-1840

Best average speed of horse drawn coaches
and sailing ships was 10 m.p.h

1850-1930

Steam locomotives averaged 65 m.p.h
steam ships averaged 36 m.p.h

1950s

Propeller aircraft
300-400 m.p.h

1960s

Jet passenger aircraft
500-700 m.p.h

Image 13.2 As transportation technology has improved, the amount of time it takes to travel long distances has decreased. We call this process the time–space compression.

Harvey, David. *The Condition of Postmodernity: An Enquiry into the Origins of Cultural Change.* John Wiley and Sons. 1992

As the energy economy changes, more transportation likely will be taken over by electricity, despite the current dominance of fossil fuels, especially petroleum.

Cars and Trucks

Fossil fuel–based transportation began with ships and trains, which greatly reduced travel times for goods and people in the nineteenth century. While train and ship transport are both important today, the transportation method that defined the twentieth century, the automobile, has shaped almost every aspect of modern society (Image 13.4). Cities have been designed and redesigned with cars in mind, vast spaces have been paved over to build roads, highways, and parking lots, and in most industrialized countries a family owns at least one personal car, while millions more buses, taxis, and other vehicles are used for various transit needs and desires.

Image 13.3 Changes in how we get energy for transportation will have major impacts in the future of transit. Today, petroleum products dominate transportation fuels.
Maksim Vivtsaruk / Shutterstock

Image 13.4 The development of the personal car with a gasoline engine helped make oil the dominant energy resource in the twentieth century.
Harris & Ewing, Library of Congress

Cars have so thoroughly shaped modern society that, especially in the United States, we are essentially dependent on them. While urbanization and development could place more importance on public transport and high-speed rail in the coming decades, cars will likely remain the primary mode of personal transportation for many years to come.

Steam power was the first major transportation use for fossil fuels. However, in the early years of automobile development, steam and electric batteries lost out to the **internal combustion engine** (Image 13.5). The internal combustion engine, first commercialized in France in 1859, represented an important leap in propulsion technology; an internal combustion engine is much smaller than a steam engine and only requires one fuel (gasoline or diesel in most cases) rather than water and a combustible material. In 1886, Daimler-Benz built the first automobile to be powered by an internal combustion engine, kicking off the development of the modern car.

Cars, trucks, and motorcycles have become some of the largest users of energy in the industrialized world, and the share of their energy appetite grows as large countries such as China and India develop at a rapid pace, opening opportunities for more people to own their own personal transportation—a hallmark of life for many in developed countries. Transportation as a whole consumed 28 percent of the United States' entire energy budget in 2013, the majority of which was consumed by the automotive sector. The energy needs and challenges of cars differ from those of other energy

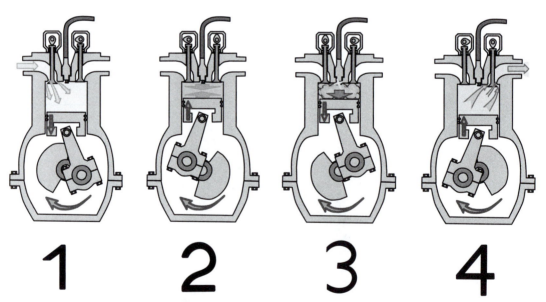

Image 13.5 An internal combustion engine takes in fuel (such as gasoline), where it is then compressed and ignited, and the waste gases are expelled. The process drives a piston that produces power, which can be used to move a vehicle, among other applications.
Sergey Merkulov/ Shutterstock

consumers because cars themselves are mostly privately owned, while most of the infrastructure they travel on (fueling stations aside) is publicly owned and funded. Consequently, efforts to improve efficiency or shift to more technologically advanced energy sources require the new technologies to be affordable and convenient, increasing the barriers to early adopters. This has been particularly apparent with electric cars, which are only now beginning to see any sort of mass production. Despite these challenges, new energy automobile choices like electric and fuel cell cars have recently begun challenging the dominance of fossil-fueled automobiles, spurred by concerns with pollution, efficiency, safety, and the dangers to our economy from unstable fossil fuel prices (Image 13.6).

Conventional fuels (gasoline and diesel fuel) were used in the earliest internal combustion engines and continue to dominate the automotive sector's energy use. Gasoline alone represents about 56 percent of the United States' transportation energy budget, with diesel fuel accounting for another 22 percent (Image 13.7). Biofuels and other alternative fuel sources have grown rapidly, but conventional fuels remain the default for personal and commercial vehicles, with biofuels blended in as a small percentage of the total (few vehicles in the United States run solely on

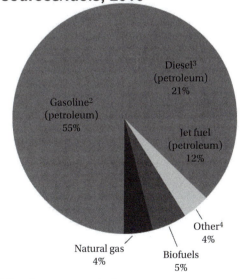

U.S. transportation energy sources/fuels, 2016[1]

Gasoline[2] (petroleum) 55%

Diesel[3] (petroleum) 21%

Jet fuel (petroleum) 12%

Natural gas 4%

Biofuels 5%

Other[4] 4%

Image 13.6 Gasoline dominates U.S. transportation fuels, along with other petroleum derivates such as diesel and jet fuel. Together, crude oil–derived products account for the majority of energy in the U.S. transportation system and globally as well.

Courtesy of the U.S. Energy Information Administration

[1] Based on energy content
[2] Motor gasoline and aviation gas; excludes ethanol
[3] Excludes biodiesel
[4] Electricity, liquefied petroleum gas, lubricants, residual fuel oil, and other fuels

Note: Sum of individual components may not equal 100% because of independent rounding.

eia

Transportation energy use by type

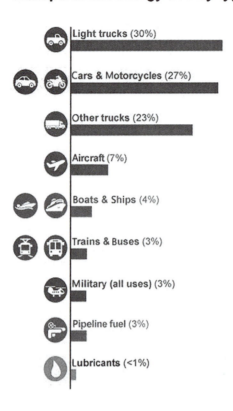

Light trucks (30%)

Cars & Motorcycles (27%)

Other trucks (23%)

Aircraft (7%)

Boats & Ships (4%)

Trains & Buses (3%)

Military (all uses) (3%)

Pipeline fuel (3%)

Lubricants (<1%)

Image 13.7 Cars and trucks use the majority of energy consumed in the United States for transportation.

biofuels, whereas over half of the automobile fleet in Brazil runs on ethanol). Improvements in efficiency have increased the longevity of use for conventional fuel vehicles; in 1975, the average fuel economy of a noncommercial vehicle in the United States was 13.1 miles per gallon (mpg) (18 liters per 100 kilometer), whereas in 2013, it had improved to 24 mpg (9.8 liters per 100 kilometer). Beyond averages, cars sold specifically for fuel economy can exceed 50 mpg averages (4.7 liters per 100 kilometer), and federal mandates continue to push the overall efficiency upward. The overall mileage traveled by cars in the United States, and indeed the total number of vehicles on the roads, has continued to grow, but efficiency improvements have led to total energy consumption by the transportation sector remaining relatively stable since around 2007.

Diesel vehicles have been less prevalent than gasoline vehicles in the United States, but diesel fuel's advantages have allowed it to remain popular for certain applications. In the United States, diesel is mostly consumed by large trucks, buses, and freight vehicles because **diesel engines**, although they cost more to buy, are more fuel efficient than gasoline engines. Diesel technology in cars has long been favored in Europe, where the higher fuel efficiency of small diesel engines has kept pace with high gasoline taxes and limitations on engine size. Many small turbocharged diesel engines achieve fuel efficiency levels comparable to that of gas–electric hybrid vehicles, but without the cost of the hybrid electric system. Compared to gasoline engines, however, diesel engines also produce higher particulate emissions, particularly of nitrous oxide compounds. With the limited availability of diesel engines in the United States and the rise of hybrid and electric vehicles, it appears for now that diesel will not catch on at the level it has in Europe, although it still represents a more fuel-efficient option for conventionally powered cars and trucks. However, some diesel studies appear to overstate the efficiency gain, as demonstrated in 2015 when a scandal broke out regarding the cleanliness and efficiency of Volkswagen diesel cars, which were programmed to cheat while having their emissions tested. This scandal is likely to damage the adoption of diesel vehicles and harm the claims that diesel is environmentally friendly.

Gasoline and diesel are currently the clear leaders of automotive transportation, but the development of true energy alternatives from hybrid to electric to fuel cell cars in the past few years has presented the first plausible challenge to the dominance of fossil fuels in the field of transportation. For an alternative energy source in automotive transportation to be truly viable, it must overcome several hurdles with convenience (must have sufficient range, must have appropriate infrastructure in place) and cost (must be affordable to most of the public, either with or without governmental incentives) being the most apparent.

Hybrid cars have been the first alternative to break through and enter mainstream use (Image 13.8). By effectively combining recent advances in electric car

Image **13.8** The Toyota Prius is one of the most successful hybrid gasoline–electric cars.
Jose Gil / Shutterstock

propulsion with the convenience of a gasoline engine, hybrids have allowed electric and battery technologies to be developed at a consumer scale without requiring major overhauls in infrastructure and manufacturing. Despite their increased efficiency, hybrids are not a full alternative because they still use fossil fuels as their primary input; the onboard battery technology only serves to increase the efficiency of the conventional gasoline engine and gains additional power from braking and rolling downhill. **Plug-in hybrids** take the hybrid engine concept one step further, by allowing a battery that is usually bigger than the battery in a regular hybrid to be charged directly from grid power while parked and then run in electric-only mode for some number of miles (Image 13.9). The onboard gasoline engine then is only required to charge the battery systems, and shorter trips can be made without any gasoline use.

Fully electric cars have been in off-and-on production since the early 1900s, but have only recently begun to make inroads into mainstream adoption (Image 13.10). Traditionally, electric cars have been disadvantaged by heavy batteries, high cost, and short range. With significant advances in battery and computer technology pioneered by personal electronics such as cell phones and laptop computers, electric cars can now be competitive as these technologies are scaled up. **Range anxiety** (the concern that an electric car's range is limiting, even when most trips are short, based on both distance and time for recharge) remains a barrier to the wider adoption of electric cars because most commercially available models have ranges under 100 miles (161 kilometers) on a single charge. Companies such as Tesla have made a significant impact on the electric vehicle market by offering luxury models with ranges that rival that of conventional gasoline vehicles, and upcoming electric cars from other companies are aiming to reduce range anxiety by offering similarly competitive ranges of over 200 miles (322 kilometers).

Image 13.9 The Chevy Volt uses a plug-in hybrid system.
Roman Vukolov / Shutterstock

Image 13.10 Teslas are some of the most successful modern electric-only cars.

Taina Sohlman / Shutterstock

If past advances in battery technology efficiency continue, range anxiety will likely cease to be an important factor in electric car usage. Infrastructure, such as a wide network of fast charging stations (which still take many hours to fully charge most vehicles), also remains a barrier to electric car usage, and while some states are building stations at a rapid pace, coverage must be far more geographically expansive for electric cars to gain widespread appeal, especially in large countries with dispersed populations, such as the United States or Canada. The difference in battery performance by season (much shorter range in cold weather) also is a problem for many consumers. Unlike shifting to hydrogen, as is sometime mentioned for transportation, the power grid in most industrialized countries already has nearly complete geographic coverage, so building charging stations is relatively cheap and easy, with no pipelines or specialized equipment needed.

The transition away from fossil fuels for the automotive sector has gained traction in recent years, and the current pace of growth of electric, hybrid electric, and alternative fuel vehicles seems likely to continue into the future. For example, the transition to an all-electric car fleet relies on the continued progress of technology and a sustained public demand, but the benefits are also clear. Energy for electric cars could easily come from renewable sources, and a large enough electric car fleet may even allow for distributed grid storage using the batteries of electric cars. The increase in efficiency of grid power over individual gasoline and diesel engines offers a major advantage from an emissions standpoint, and the greater mechanical simplicity of electric motors (thus lowering repair costs) offers a compelling case to owners to switch to electric vehicle technology, assuming that the costs of electric vehicles can drop to a point where they are competitive with current conventional vehicles.

Aviation

Air travel was once exclusive, expensive, and out of reach for most people (Image 13.11). The development of efficient **jet engine** aircraft and pressurized airplane cabins, beginning in the 1950s, kicked off air travel as a major means of transportation for people and goods. Airlines using jet and propeller aircraft now exist in almost every country, and air travel has effectively made it possible to travel to almost any populated place on the planet in less than a day. A major consequence of the incredible growth of air transit has been a vastly increased demand for energy. While other transportation methods can run on electricity or alternative fuels, aircraft still run on essentially the same fuels that they have used since the development of the first jet engines—mostly petroleum-derived fuels.

Jet fuel accounted for 11 percent of all transportation energy usage in the United States in 2015, a figure that has been growing over time as air travel becomes cheaper and more routine. Airlines have virtually eliminated all competition from other transit modes in long-distance international travel, and the underdeveloped rail system in the United States has allowed airlines to nearly monopolize domestic long-distance travel as well, a phenomenon that is now playing out in other big countries where rail is less efficient. Because of the overwhelming advantages of speed and time savings, air travel has continued to grow, despite its gluttonous energy appetite. Fuel costs made up around 25 percent of all expenses for airlines in the United States in 2015, a figure similar to that in other countries. Air travel has continued to grow in developed countries, but growth in developing economies has been rapid. China went from 266 million air passengers in 2010 to over 390 million passengers in 2014, while countries such as

Image 13.11 The Boeing 707 helped usher in the jet age, where large numbers of people and large amounts of cargo travel by air, using petroleum-derived jet fuel.

Ivan Cholakov / Shutterstock

Turkey have more than doubled their number of air passengers in the same period. This massive growth in air travel has resulted in a commensurate rise in energy use in the sector, and air travel is forecasted to continue its rapid growth for many years. Because there are currently no viable alternative energy sources for passenger air travel, the air travel sector will likely remain a major consumer of fossil fuels (and, to a lesser extent, biofuels) for years to come.

Air cargo has gained a major foothold in the global economic system at the same time as passenger air travel, although it is among the least efficient methods of transport available. While trucks, trains, and ships carry the vast majority of the world's cargo, air transport has enabled logistics corporations to ship packages anywhere in the world in a matter of hours, and perishable commodities such as flowers and seafood have become available almost anywhere because of air transit. As a major example, FedEx is the world's largest airline, despite not carrying passengers. Its primary hub in Memphis, Tennessee, is also the United States' busiest cargo airport (second only to Hong Kong in the world), having transported over twenty-one billion pounds of cargo in 2014. Despite the massive efficiency advantage of other forms of transport, air cargo has greatly shrunk the world for several economic sectors that rely on short delivery times.

Alternative energy technologies have begun to compete seriously in the automotive and rail sectors, but aviation presents a number of challenges to adopting electric or alternative power sources. For aircraft design, weight is a crucial factor; consequently, the high energy density of refined fossil fuels like jet fuel is ideal because a plane can carry enough on board to provide intercontinental range. Many airlines have begun to experiment with biofuels made from a variety of sources, including used cooking oil and waste, as a way to reduce their petroleum consumption. Cathay Pacific Airways and United Airlines are among the leaders of this movement, purchasing equity stakes in biofuel producers and committing to using millions of gallons of biofuel annually. However, this still represents a drop in the bucket of their annual fossil fuel consumption; for example, United's commitment to use ninety million gallons of biofuel over ten years could only fuel its 2015 flight schedule for about four days. Airline interest in biofuels is unlikely to drive a significant shift away from petroleum products for many years to come.

At present, electric propulsion systems have far lower energy density than fossil fuels and have only been used to power small, lightweight experimental aircraft. In the long term, viable electric propulsion systems for commercial aircraft may become available, but have been much slower to develop than electric systems for automobiles. Much of the drive for energy efficiency in the aircraft sector has been driven by weight saving and engine efficiency, rather than fuel type. Large modern aircraft such as the Boeing 787 or the Airbus A380 achieve much higher fuel efficiency by drastically reducing the weight of their aircraft, using modern materials like carbon fiber and strong aluminum alloys (Image 13.12). With better engines, light airframes, and higher passenger capacities, modern aircraft are far more efficient than their predecessors, but they still use a large part of the total transportation energy budget in the United States and globally.

Image 13.12 The Airbus A380 is the largest passenger jet ever manufactured.
Paul Drabot / Shutterstock

Rail

Rail transport has played a defining role in the modern history of humanity, from opening the western United States and Canada to concentrated settlement to the high-speed bullet trains of Japan and France. Railroads are a major investment of infrastructure and equipment, but are also one of the most efficient methods of overland transport available. Despite the energy advantages of rail transport for many economic sectors, rail use and development vary widely among different countries, states, and cities.

Before the automobile became ubiquitous in the twentieth century, trains were the most important land-based transportation in industrialized countries (Image 13.13). Even today, trains are crucial for freight and passenger transport, primarily because of their high energy efficiency. Electrified train networks, such as those used in Europe, Japan, and on the U.S. East Coast, take electricity directly from the main grid and use it to power rail systems, so their energy efficiency is directly tied to the original generation source. For rail systems without electrification, such as much of the U.S. network, diesel has been the primary fuel. While diesel trains are less environmentally friendly than trains run by direct grid power, they still manage an average of four times higher fuel efficiency than highway-based freight, with some lines claiming that a gallon of fuel can move a ton of cargo almost 500 miles (800 kilometers). Trains can achieve such high efficiency through scale because most of their momentum is conserved by lower rolling resistance, the fixed nature of tracks, and the limited need to stop between destinations.

Passenger train use is similarly energy efficient, yet its use varies substantially across different societies. The relative dominance of cars and aircraft in the United States has relegated passenger trains to a minor role in long-distance travel. For instance, Amtrak carried 30,921,274 passengers in 2014, while airlines in the United

Image 13.13 Railroads once dominated intercity travel and remain important in many parts of the world for passenger travel.
Jack Delano, U.S. Farm Security Administration, Library of Congress

States carried 726,560,000 passengers in the same year (this equals an average of three flights per person)—almost twenty-five times as many. Because of the cost and time for long-distance rail travel, Amtrak primarily serves customers in dense urban corridors, particularly that between Boston and Washington, DC (Image 13.14). The only high-speed rail system in operation in the United States is the Acela, run by Amtrak, which runs between Boston and Washington, DC, at a maximum speed of 150 mph (241 kph) (Image 13.15). Although the Acela network is relatively small, it earns nearly 25 percent of the total Amtrak revenue every year because of its popularity and convenience in regional transport. Proposals for similar high-speed rail systems in other parts of the United States have been suggested as a more energy-efficient solution over air travel, although the large infrastructural costs and limited success of rail outside of the eastern United States have so far prevented any major new networks from being constructed anywhere but California, where a major system should be complete by 2029.

While long-distance and regional rail systems have met with varying success in the United States, rail in urban centers as a commuting system remains popular. The New York City subway system is the highest profile metropolitan rail system in the country (Image 13.16), with over 1.75 billion rides in 2014, although most major cities have adopted some form of subway, light rail, or commuter rail network. In recent years, public referenda and new political support for transit have led to new construction on rail systems in many cities, with the hope that such systems will reduce traffic and be more environmentally sustainable. Despite this growth, New York City accounts for one-third of all mass transit riders in the United States and a whopping two-thirds of those that use rail (compared to buses). This massive ridership only accounts for 30 percent of commutes in New York, and the next highest cities—Washington, DC, and San Francisco—have fewer than 15 percent of their commutes via mass transit. In parts of East Asia and Europe, the percentage of commuters via rail is higher, especially in major cities such as London, Tokyo, and Moscow. In North America, New York is second to Mexico City, where mass transit accounts for 70 percent of commutes, many via its massive subway system.

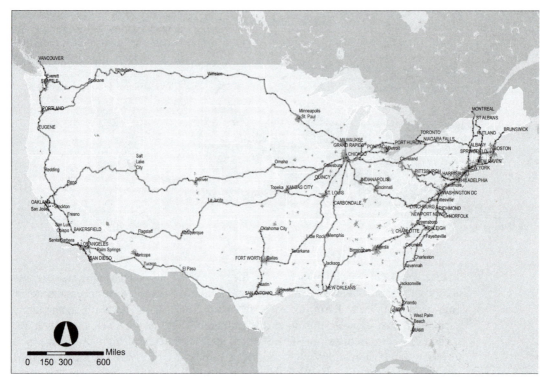

Image 13.14 Amtrak is the passenger rail system for the United States.
User:Pi.1415926535 / Wikimedia Commons / CC-BY-SA-3.0

Image 13.15 The Acela is the high-speed rail system developed for the United States. New track improvements and some changes to routes will be required to allow it to go as fast as high-speed rail systems in Asia and Europe.

Lee Snider Photo Images / Shutterstock

Image 13.16 Subways provide a heavy-rail mass-transit option in dense cities. They use electricity, rather than liquid fuels.
Natalia Bratslavsky / Shutterstock

Shipping

Nautical transportation has been crucial for human societies for thousands of years and in many ways continues to be the most efficient form of transportation available. Cargo ships in particular are not constrained in size in the same way as trains, automobiles, or aircraft, so they achieve unparalleled efficiency by utilizing massive economies of scale and taking advantage of the relative lack of friction of water versus land travel. The world's largest container ship, the MSC *Oscar*, is almost 400 meters (1,312 feet) in length and is capable of carrying nearly twenty-thousand twenty-foot shipping containers (Image 13.17). The biggest trains can carry at most four hundred shipping containers. Even with conventional propulsion, ships like the *Oscar* are far more fuel efficient than anything else available. While the low speed of large ships cannot compete with that of aircraft or automobiles for passenger transport outside specific roles, the globalized economy relies on the efficiency of shipping to maintain the massive flows of cargo between countries and continents. However, these large ships often run on dirty and cheap fuel derivatives such as heavy or bunker oil, which is often high in sulfur, leading to significant air pollution problems.

Because of their low speeds and high weight capacities, ships have long been adaptable to new propulsion technologies. From wind to steam to modern diesel turbines, almost every type of energy source has been used in shipping at some point. Perhaps the most prominent example of alternative energy technology in ships is the use of portable nuclear reactors. Nuclear-powered shipping has most prominently been used to power aircraft carriers and submarines (Image 13.18), but has also been used in the Arctic to power icebreaker ships. Nuclear ships have a distinct range advantage because they require refueling every few months or even years, but the cost of construction, in

Image 13.17 The MSC *Oscar* is the world's largest container ship.
PtnPhoto / Shutterstock

Image 13.18 Only military craft, such as U.S. aircraft carriers, use nuclear power for propulsion.
Everett Historical / Shutterstock

addition to concerns of safety and fuel disposal, have meant that nuclear power remains rare for nautical applications except for the military. With modern advances in alternative energy and battery technology, however, cargo ships may soon be able to go completely renewable. Electric power is well suited to nautical applications, so solar-powered shipping may one day be used for cargo for which quick delivery is less important. Automation technology may even make it possible for low-speed solar ships to be completely uncrewed, which would reduce costs and allow ships to travel at slower, more energy-efficient paces without needing to worry about crew comfort.

Electric Vehicles When the Electricity Comes from Fossil Fuels

If the electricity for an electric vehicle comes from solar- or wind-generated electricity, those electric miles are sustainable. But what about where the electricity comes from coal, the dirtiest fossil fuel with the highest carbon dioxide emissions of any fossil fuel (Image 13.19)? And what about the emissions and pollution associated with manufacturing electric cars and batteries? Factoring in everything from manufacturing, operation, and disposal of an electric vehicle at its end of life, electric vehicle are better than gasoline cars in every state of the United States. The Union of Concerned Scientists published a report showing that although the emissions from manufacturing electric vehicles are higher than that from gasoline cars, in six to eighteen months of operation, those extra emissions are offset by lower operational emissions, even when including carbon dioxide emissions from the electric grid. The report goes on to show that to match the total emissions from electric vehicles, gasoline cars would need to average better than sixty-eight miles per gallon when compared to the average emissions of electricity generation across the United States.

Image 13.19 When the electrical grid is powered by fossil fuels, like coal, a plug-in electric car is still deriving its energy from fossil fuels.
Nui7711 / Shutterstock

Technological Advances in Transport

As with other energy systems, technological developments have the capability to completely revolutionize the way we consume energy for transportation. While many of the most major advances in recent years have been in the development of hybrid and electric drive systems, hydrogen and other alternative fuels could potentially make up a major part of transportation fuel usage in the future, provided that the technology to produce them is sufficiently advanced. Obstacles to the use of biofuels and hydrogen (discussed in Chapters 10 and 12) remain significant, but major investments in infrastructure and research could allow either of these fields to grow beyond their small roles today and reshape the future of transportation.

REVIEW QUESTIONS

1. Why is transportation an important energy issue today? How is a switch to renewable electricity harder to implement here than in other sectors? Is there a difference in how this is addressed for different forms of transportation?
2. Why is aviation likely to consume liquid fuels long after other transportation sectors convert to electricity?
3. How can electricity play a bigger role in city commuting (via both personal and mass transit)?
4. Why must we focus not only on human transport, but also on goods?
5. Are there additional challenges with aviation and shipping?

Image 14.1 Trofimov Pavel / Shutterstock

CHAPTER 14

Air and Water Pollution

Pollution of both our air and our water is a central challenge for the sustainability and viability of human energy consumption. While fossil fuels remain central regarding the impact of energy production and consumption on our air and water, other energy resources also affect our environment. The specific role of pollutants in climate change has become so central to our current discourse and politics that they are addressed separately in the next chapter.

Air and water pollution have both been energy-related challenges since the advent of the fossil fuel age. The famous London fog was a direct result of coal burning, just as Los Angeles' famous smog is directly tied to the cars that clog its infamous freeways. Many of the pollutants that flow through our water also derive from energy production, transportation, and consumption.

Because of these links between energy and environmental pollutants, much of the regulation of energy production and consumption is directly tied to our efforts to control this pollution. The landmark Clean Air and Clean Water Acts, passed by the U.S. Congress and signed into law in 1963 and 1972, respectively, serve as the primary basis for many of the regulations that control energy production in the United States. Prior to these laws, little of the pollution produced by energy use, especially via fossil fuel use, faced any sort of legal restriction. We will investigate the various air and water pollutants derived from energy resources and how we address them, with a focus primarily, but not exclusively, on fossil fuels.

Air Pollution

The advent of the fossil fuel age made the atmosphere one of humanity's principal dumping grounds. Although the burning of wood had already generated polluted air, the use of coal and later other fossil fuels greatly increased air pollution, both in local areas and globally. Early on, this was seen with the famous London fog, which was actually smog that blanketed the city from the widespread use of coal to heat homes. In general, **air pollution** is the release of any damaging substance into the earth's atmosphere, while **ambient air pollution** refers to outdoor air pollution found mostly in areas where humans burn fuels or waste and otherwise release chemicals into the air.

Most of the air pollution we deal with stems from human sources, led by energy resource production and consumption, but we must remember that natural air pollution also exists. The many sources include volcanic eruptions, dust storms, and naturally occurring fires. Non–energy related human-caused pollutants include industrial chemicals such as **chlorofluorocarbons**, which deplete the ozone layer, among others.

Energy-related air pollutants, however, form a major component of current global air pollution, primarily driven by the burning of fossil fuels. One of these, carbon dioxide, will be discussed separately in Chapter 15 on climate change, as will the climate-specific impacts of other air pollutants such as methane. In the United States, six air pollutants are tracked as the so-called **criteria pollutants**, because they are the most dangerous for humans (Image 14.2). These pollutants are as follows:

- **Sulfur dioxide**: Sulfur dioxide is primarily produced by the burning of sulfur-bearing coals for the production of electricity and in industry, as well as in unrefined petroleum. Sulfur combines with oxygen during combustion to produce this gas. Sulfur dioxide pollution is the primary driver of acid rain (discussed below in the section on water pollution) and is harmful in and of itself. When inhaled in large amounts, sulfur dioxide can lead to asthma and other respiratory conditions.
- **Nitrous oxides**: Nitrous oxides, nitric oxide and nitrogen dioxide, are produced from the heat in combustion engines, primarily in motor vehicles. Some is also produced in power plants and factories. These chemicals are highly reactive and lead to acid rain and smog production.
- **Ozone**: Ozone in the stratosphere protects life from ultraviolent radiation; however, human-produced smog in the troposphere (the lowest level of the atmosphere) is a pollutant. Ozone is highly reactive with sunlight, nitrous oxides, volatile organic compounds, and heat and is one of the primary drivers of smog.
- **Carbon monoxide**: This chemical is produced from incomplete combustion mostly in transportation engines (78 percent of carbon monoxide in the United States is produced this way), but also from other combustion of fossil fuels, including home heaters and wood. It is significantly dangerous, even in small amounts, and can be deadly because it binds to hemoglobin in the blood and prevents the body from properly absorbing oxygen.
- **Particulates**: This general category of air pollutants consists of minute liquid and solid particles suspended in the atmosphere, such as soot and dust. These particulates can be inhaled and damage lungs. Most particulate pollution is

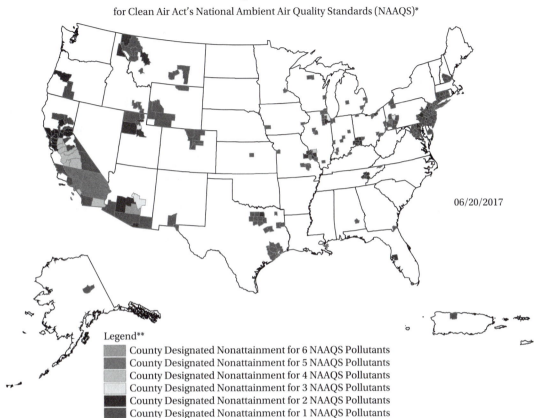

Counties Designated "Nonattainment"
for Clean Air Act's National Ambient Air Quality Standards (NAAQS)*

06/20/2017

Legend**

☐ County Designated Nonattainment for 6 NAAQS Pollutants
☐ County Designated Nonattainment for 5 NAAQS Pollutants
☐ County Designated Nonattainment for 4 NAAQS Pollutants
☐ County Designated Nonattainment for 3 NAAQS Pollutants
☐ County Designated Nonattainment for 2 NAAQS Pollutants
☐ County Designated Nonattainment for 1 NAAQS Pollutants

Guam-Pitiand Tanguisson Counties are designated nonattainment for the SO2 NAAQS

*The National Ambient Air Quality Standards (NAAQS) are health standards for Carbon Monoxide, Lead (1978 and 2008), Nitrogen Dioxide, 8-hour Ozone (2008), Particulate Matter (PM-10 and PM-2.5 (1997, 2006 and 2012), and Sulfur Dioxide. (1971 and 2010)

**Included in the counts are counties designated for NAAQS and revised NAAQS pollutants. Revoked 1-hour (1979) and 8-hour Ozone (1997) are excluded. Partial counties, those with part of the county designated nonattainment and partattainment, are shown as full counties on the map.

Image 14.2 Air pollution is most concentrated in urban areas where large numbers of vehicles, and in some cases coal-fired power plants, exist.

Courtesy of the U.S. Environmental Protection Agency

naturally occurring, but some is driven by the burning of coal and is a major problem in China because of the country's enormous coal consumption.

- Atmospheric **lead**: Lead is a toxic substance that is especially dangerous to young children because it retards mental development, and it is also a carcinogen. Since the 1970s, tetramethyl lead has been banned in the United States and other developed countries, but in many developing countries it remains an additive to gasoline that improves engine performance.

Beyond the six criteria pollutants, other air pollutants are also major problems, including the **volatile organic compounds**. Many of these are hydrocarbons released on their own into the atmosphere, including the simple natural gases—methane, ethane, propane, and butane. Along with several of the criteria pollutants, volatile organic compounds can be a driver for smog and are also dangerous when inhaled into the body. Aerosols are another component of air pollution traceable to fossil fuel use. These are fine droplets of condensed air pollutant, many of which reflect sunlight back into space and can cause cooling. Most aerosol pollution derives from other sources, but some is produced during combustion of fuels. Finally, **mercury** can also be released during coal combustion. This heavy metal causes similar problems to lead and is a dangerous neurotoxin.

Although not a pollutant itself, **smog** is the visible byproduct of various forms of energy-related air pollution. The term smog means smoky fog and is the colloquial term for a class of ambient air pollution that especially impacts urban areas. Smog is caused by the reaction of volatile organic compounds and/or nitrous oxides with sunlight to produce a hazy layer of air pollution. It often contains smoke and sulfur dioxide as well. Smog can be caused by burning coal to produce electricity or for industry (**industrial smog**) or from motor vehicles (**photochemical smog**). In some areas, smog concentrates by thermal inversion in the atmosphere that traps these pollutants near the ground. Los Angeles, Beijing, and Mexico City are especially noteworthy for this (Image 14.3); however, New Delhi, India, is now likely the city with the worst smog. Smog is especially harmful to children and the elderly, along with anyone with asthma or other respiratory problems. Smog can lead to premature death and makes outdoor activity dangerous.

Industrial smog remains a major problem in developing countries, especially China and India. The smog of the past in the developed world was also mainly of this type. Photochemical smog remains a major problem in developed countries, as well as in booming developing-world cities where more cars are on the roads today.

Image 14.3 The smog of Los Angeles is caused by the large number of vehicles and the local topography, which traps pollutants.
imantsu / Shutterstock

Over time, developed countries have made significant improvements in air pollution. In the United States, this improvement has been driven by the adoption of the **Clean Air Act** in 1970. The Clean Air Act has enabled the U.S. **Environmental Protection Agency** to both monitor and enforce reductions in air pollution. Overall air pollution levels in the United States have fallen by over half, with lead by over 99 percent and particulates by 85 percent. Over the same time period, the economy has grown, vehicles have traveled further, and energy consumption has increased, demonstrating that pollution limits do not inhibit growth. However, despite these massive reductions, air pollution still causes an estimated 4 percent of annual deaths in the United States (Image 14.4).

Several technological breakthroughs have aided the decrease in air pollution. Motor vehicles are now equipped with **catalytic converters** (Image 14.5). These converters treat vehicle exhaust by passing it through a filter where aluminum oxide, rhodium, palladium, and platinum serve as catalysts to chemical reactions in which harmful gases are converted into carbon dioxide, water vapor, and nitrogen.

Another major development has been **scrubbers**, which have significantly reduced air pollution releases from coal-fired power plants (Image 14.6). Scrubbers use a variety of processes to remove pollutants, both via filtering them out and by converting others in chemical reactions into harmless substances. Flue gas passes through a scrubber prior to release into the atmosphere (Image 14.7). Coal power plants release large amounts of **fly ash**, a residue of fine particulates, which are largely filtered out by scrubbers. Chemically treated water is mixed with the gas to remove particulates, sulfur, nitrous oxides, and other harmful compounds contained in the ash. The water can then be settled and treated as hazardous liquid waste, while only certain gases, including carbon dioxide and water vapor, are released through the smokestack.

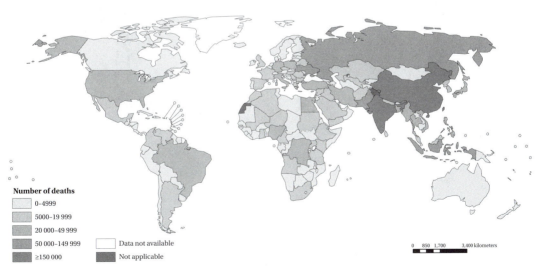

Number of deaths

	0–4999
	5000–19 999
	20 000–49 999
	50 000–149 999
	≥150 000

Data not available

Not applicable

0   850 1,700   3,400 kilometers

Image 14.4 Urban air pollution impacts millions of people around the world, with countries in Asia suffering the worst urban air pollution.

Courtesy of the World Health Organization

Image 14.5 A catalytic converter helps reduce dangerous emissions from automobile exhaust.
Einar Muoni / Shutterstock

Image 14.6 Scrubbers remove many harmful chemicals and particulates from the exhaust of power plants, especially those that burn coal.
Mark Winfrey / Shutterstock

Indoor air pollution is also a problem that can be associated with energy resources. The United Nations estimates that thousands of people die daily, and over two million per year die from indoor air pollution. The primary energy-related indoor air pollution is driven by the use of firewood or other similar materials for heating and cooking, rather than electricity or natural gas, the primary fuel sources in developed countries. Billions of people still rely on wood, crop waste, or animal dung to cook with and warm their homes (Image 14.8). This leads to large amounts of soot and carbon monoxide in the house, with levels rising often to twenty times the safe amount, according to the World Health Organization. This primarily affects the poorest and least educated people in the world who have no access to improved forms of energy to meet their basic daily needs.

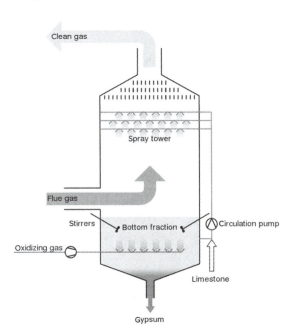

Image 14.7 Wet scrubbers use water and other chemicals to remove pollutants, such as sulfur dioxide, from flue gas.

User:Sponk / Wikimedia Commons / CC-BY-SA-3.0

Image 14.8 Burning wood or other organic fuels indoors leads to many respiratory problems in the developing world.

Pedro J Rico Lopez / Shutterstock

Water Pollution

Although many of the most pronounced impacts from energy-related pollution affect the air, water is also damaged by energy resource extraction and consumption.

Acid rain can be thought of as both an air and a water pollutant. Sulfur dioxide and nitrous oxides are the primary culprits in acid rain because they combine with water vapor, oxygen, and other chemicals in the atmosphere to form acids, particularly sulfuric acid. These acids then precipitate back to the ground, primarily as rain, but also in ice and snow, and some are dry deposited. These acids primarily derive from electricity generation via fossil fuels and from motor vehicle exhaust, although some come from non–energy related processes. Acid rain leaches nutrients from the soil and kills plants on land and in lakes and streams (Images 14.9 and 14.10). It can also kill fish and other aquatic life.

Coal-fired power plants produce large quantities of fly ash, which was often vented into the atmosphere prior to pollution controls, but now is largely stored as a solid waste product. To prevent the large quantities of ash from blowing away and escaping into the air, ash is often stored as a liquid slurry. Fly ash contains several harmful compounds, including arsenic, lead, mercury, and many other heavy metals, which have serious health risks when leached into the water supply. Because of lax rules on fly ash storage and disposal, much of the ash from coal power plants ends up leaking into groundwater. The Environmental Protection Agency estimates that 72 percent of all toxic **water pollution** in the United States comes from coal power plant runoff. Some major individual incidents have had even greater effects, such as the 2008 Tennessee Valley Kingston spill, in which 5.4 million cubic yards of fly ash were released into local rivers (Image 14.11).

Image 14.9 Acid rain, caused by sulfur dioxide emissions, impacts both plant life and aquatic environments.

Anticiclo / Shutterstock

Image 14.10 Coal ash and other solid waste can spill into aquatic environments, causing major water pollution problems.

Courtesy of the U.S. Department of the Interior, USGS Water Science

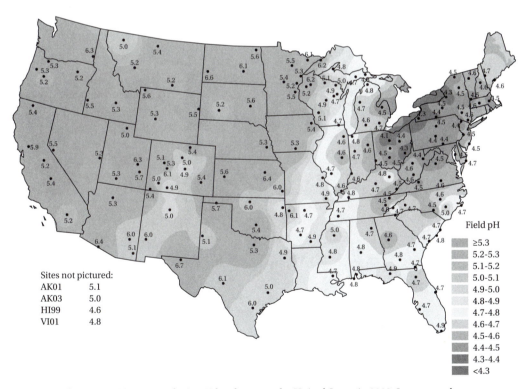

Image 14.11 A map of rain pH levels across the United States in 2002. Lower numbers are more acidic, indicating more acid rain presence.

Wade Payne / Associated Press

Mines also produce significant amounts of pollutants, many of which can eventually reach the water supply. Much like fly ash, particulate waste and leftover rock from mining, known as **tailings**, are often stored in large piles or mixed with water in storage ponds, similar to fly ash containment. These tailings contain heavy metals, mercury, arsenic, selenium, and even sometimes acids that have been used to dissolve rock materials in the mining process. When these storage systems fail, large quantities of these toxic materials end up in rivers, lakes, and groundwater. Derelict mines that have been abandoned present a particular problem because no owner is responsible for cleanup and mitigation, sometimes allowing water contamination to continue until the government steps in. In 2015, the Gold King mine in Silverton, Colorado, an abandoned mine where the Environmental Protection Agency was attempting to clean up tailing ponds, had a containment breach. In the spill, three million gallons of contaminated water spilled into the local river system, and cleanup could take many months or even years, with serious damage to local fish, wildlife, and people living in the area.

During oil drilling, pipeline transportation, and storage, petroleum products often leak into surrounding water systems, both surface water deposits (lakes, rivers, etc.) and groundwater (Image 14.13). This **hydrocarbon runoff** is toxic to humans and other animals and plants. Oil storage, especially at gas stations and other end use points, is a major water pollutant in the United States. Cars and other vehicles also leak gasoline directly into water systems.

Image 14.12 Mine tailings can leach acidic compounds into the water system.

Image 14.13 Oil leaks from rail transport, pipelines, and storage facilities contribute to water pollution problems.
Arvydas Kniuksta / Shutterstock

All thermal power plants, whether fossil fuel or nuclear, produce massive amounts of heat that are released by the plant as hot water **thermal runoff**. Warmer water has a lower oxygen content, which reduces the ability of a water body to support life, everything from microorganisms to plants to animals. Heat also leads to faster growth of plants and especially algae, which further reduces the water's oxygen content, thereby impacting life.

A similar and opposite effect takes place in many hydroelectric plants. Because the penstock is located well below the surface in colder waters, hydroelectric dams often release water downstream that is far colder than what naturally flows through the river. This colder water impacts plants and animals that are adapted for specific temperatures and has caused massive die-offs of species adapted to warmer water.

REVIEW QUESTIONS

1. What is smog, and how does it differ from other classes of pollutants?
2. What technological advances have contributed to reducing air pollution?
3. Temperature can be a pollutant in its own right. How do temperature changes impact natural environments?

Image 15.1 Trofimov Pavel / Shutterstock

CHAPTER 15

Climate Change

Energy resources play an essential role in virtually every aspect of modern society, but our appetite for energy has been one of the most disruptive forces impacting the global environment. At the same time that humanity has discovered the new interconnections in the globalized economy, we have realized that our consuming of natural resources, fossil fuels, and energy in general have truly global impacts, from rising global average temperatures, to ozone depletion, to ocean acidification, to melting ice sheets. Climate change has been a subject of rigorous scientific study for decades, but has only begun to receive massive public attention in the past few decades, especially following the Rio Earth Summit in 1992. As more and more scientists recognize the threat of climate change—and the inherent role that our energy use plays in it—a stronger focus on re-thinking our entire global energy system has developed. Despite the overwhelming scientific consensus that climate change is real, progressing rapidly, and **anthropogenic** (or human caused) in nature, many people and organizations still reject the premise of climate science. A nuanced understanding of the way humanity harvests, generates, and consumes energy requires an examination of the close relationship between our energy-intensive lifestyles and the environment.

Climate change is widely misunderstood, despite the importance it holds in the popular discourse. An important distinction when discussing climate change is the difference between weather and climate. Weather varies significantly by location and time of year and is difficult to predict with much precision. Simply put, weather is what you experience in a given place on a given day. Climate is the average of all weather for a given area over a period of time and at a particular time of year; therefore, climate is far more constant over time, whereas weather changes constantly. A warm day in mid-winter may be abnormal weather in Chicago, for example, but the climate takes variation into account. Because climate change involves changes to the long-term averages of weather patterns, it must be discussed in different terms than the weather patterns of a given, day, week, month, or even a few years. A common misconception is that certain weather events (for example, a cold and snowy winter) demonstrate that climate change is not really occurring, but the average temperatures and weather patterns in the same place do indicate warming temperatures and milder winters overall. The complexity of climate science and its various and differing effects have also largely led to a preference for the term *climate change* over the previously more common *global warming*. The changing climate has so many effects beyond temperature that warming is far from the only concern. Beyond this, some discrete areas of the planet may even experience a degree of cooling, and although these places are fewer than the warming regions, it makes climate change a far more inclusive and accurate term.

Greenhouse Gases and the Greenhouse Effect

The earth's climate is an extremely complicated system, with hundreds to thousands of interdependent factors contributing to global temperatures and weather patterns. Despite this complexity, scientists have become much more precise in their ability to model the future of our climate based on a wide range of data types. Since the changing climate was first observed by scientists in the 1960s, the science of climatology has become increasingly focused on constructing elaborate models. Early climate models were relatively simple and somewhat inaccurate by today's standards, and modern models now utilize massive amounts of computing power to incorporate thousands of different data sets to build predictive models. As observation of the planet's weather and climate system has gotten more detailed (through satellites, weather stations, and more), the volume of precise data for modeling has improved. Combined with a growing record of ice cores and tree rings that provide a climatic history of over 750,000 years, modern climatologists can develop complex mathematical models that simulate the effects of greenhouse gases (GHGs) on the global climate.

The primary driver of anthropogenic global warming is the **greenhouse effect** (Image 15.2). The earth receives most of its energy from the sun (with a small percentage coming from internal heat; see Chapter 11 for more information), and that energy is either absorbed or radiated back into space. Without an atmosphere, the planet would maintain a surface temperature of around -20°C (-4°F), but the atmosphere prevents much of the solar radiation from escaping when heat radiated from the surface interacts with atmospheric gases, instead of continuing up directly into space. The net effect of atmospheric gases on the warmth of the planet is known as **radiative forcing**, in other words, the sum of all gases that prevent excess heat from the planet from escaping, thereby contributing to excess heat being retained in the atmosphere. Greenhouse gases change the strength of radiative forcing by absorbing energy and radiating it as heat, instead of allowing energy to dissipate into space.

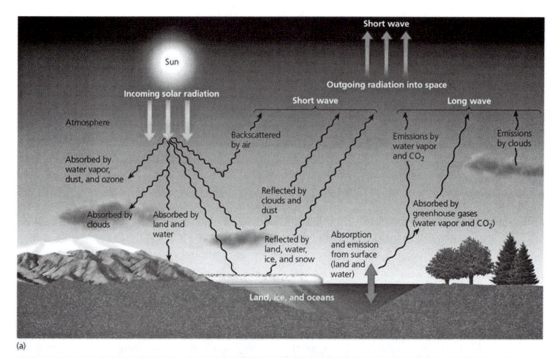

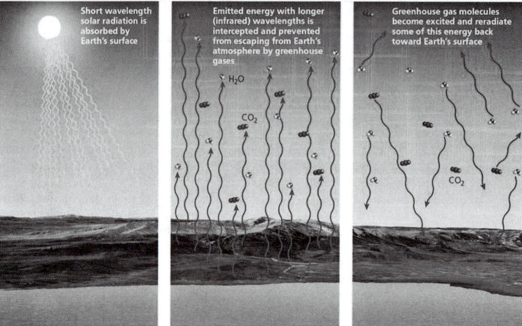

Image 15.2 (a) A large amount of solar radiation never reaches the earth's surface and is reflected back into space or absorbed by the atmosphere. (b) Shorter wavelength radiation can be absorbed by greenhouse gases, which then radiate the heat back to the surface.

Image 15.3 Emissions from power plants, such as this coal-fired facility, are an important part of human-produced greenhouse gas emissions.
Ungnoi Lookjeab / Shutterstock

There are several important GHGs, the most significant of which is carbon dioxide (CO_2). Carbon dioxide accounts for the majority of radiative forcing of the planet, although other GHGs such as methane are more potent, because there is far more CO_2 in the atmosphere and humans have released far greater amounts through the burning of fossil fuels, deforestation, and other activities. Carbon dioxide has always been a major component of the atmosphere, with concentrations varying between 180 and 290 parts per million (ppm) over the course of the climatic record. In nature, CO_2 is sequestered in organic material and is then buried in soil, a process known as **fixing**. Plants take in CO_2 and emit oxygen during photosynthesis, during which the carbon is stored as part of the plant. When the plant dies and is buried in soil or sediment, it is considered fixed, or removed from the atmosphere. By burning fossil fuels that are composed of fixed carbon from thousands to millions of years ago, excess CO_2 is emitted into the atmosphere, thereby altering the strength of radiative forcing. Compared to the historic highs of 290 ppm as revealed by ice core records, the current atmospheric concentration of CO_2 is just under 400 ppm. Virtually all of the excess CO_2 in the atmosphere is a result of human activity. Most climatologists agree that the maximum level of atmospheric CO_2 to keep atmospheric warming at or below 2°C is around 350 ppm, a level that has been exceeded since 1988.

The majority of atmospheric CO_2 emissions comes from the burning of fossil fuels. Because of continuing industrialization, the growth of personal automobile usage, and increasing electricity consumption, the rate of fossil fuel combustion has been increasing considerably worldwide for many decades. Between 1994 and 2004 alone, CO_2 emissions increased by nearly 25 percent. Changes in primary fuel sources and emissions technologies have altered which countries are the biggest emitters of CO_2, but industrialized countries still account for the overwhelming majority of the historical total of emissions since the beginning of the Industrial Revolution (Image 15.4).

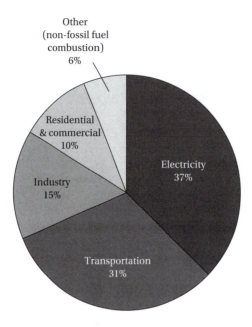

Other
(non-fossil fuel
combustion)
6%

Residential
& commercial
10%

Industry
15%

Electricity
37%

Transportation
31%

Image 15.4 Electricity production is the number one source of atmospheric carbon dioxide pollution from the United States. Courtesy of the U.S. Environmental Protection Agency

Although CO_2 is the largest contributor to radiative forcing, there are several other important GHGs. Methane, the second largest contributor to radiative forcing, is naturally emitted through bacteria and decomposition and is removed from the atmosphere through chemical reactions between ozone and solar radiation. Methane is nearly thirty times more powerful than CO_2 by mass in its effect on radiative forcing, but atmospheric concentrations are far lower. Methane in the atmosphere has increased from a historic level of 750 parts per billion (ppb) to 1750 ppb in 2005. Unlike CO_2, atmospheric methane comes from a wide variety of sources. Many anthropogenic sources of methane exist, including agriculture, natural gas and petroleum systems, landfills, mining, and deforestation. Because of increasing global population and increased development, the emissions of methane from these sources have overwhelmed the natural systems that keep atmospheric methane at stable levels.

While the growth in global methane emissions has slowed, the warming climate may introduce far more methane in the future. Permafrost, which contains a massive amount of organic material, remains relatively inactive because it does not seasonally thaw like normal soil. If warming continues to cause large amounts of permafrost to thaw, the organic material within it can then decay, releasing far more methane than current industrial processes (Image 15.5). This could cause a feedback loop, wherein further heating is caused by emitted methane, thawing yet more permafrost. The scale of this possibility is not yet fully understood, but it represents one of the many possible catastrophic consequences of a warming climate.

Although CO_2 and methane are the two most significant contributors to radiative forcing, several other gases play significant roles as well. Nitrous oxide (N_2O) is a major GHG, largely released into the atmosphere through fertilizer, industrial production, and land

Image 15.5 Areas in the polar latitudes are seeing climate change at the most rapid levels.
dinozzaver / Shutterstock

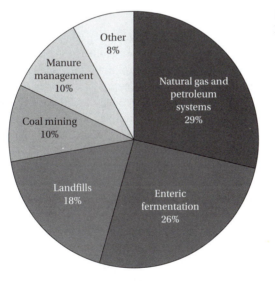

Image 15.6 Courtesy of the U.S. Environmental Protection Agency

clearance. Because nitrogen is fixed by plant growth, deforestation exacerbates the role of N_2O in warming. The majority of N_2O emissions come from agriculture, with just 5 percent coming from transportation through fossil fuel combustion (Image 15.7). Similar to methane, N_2O absorbs different wavelengths of radiation, making it a much more potent gas than CO_2, by a factor of 300. Nitrous oxide accounted for about 5 percent of U.S. GHG emissions in 2013, and 40 percent of N_2O emissions globally come from human activities. A particular concern with N_2O is that, while it is naturally removed and fixed through

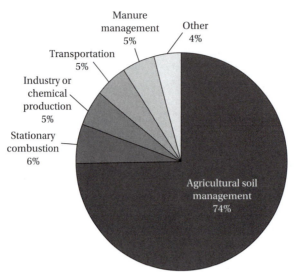

Image 15.7 Energy systems are a part of nitrous oxide emissions, but agricultural practices are the main contributor of this pollutant.

Courtesy of the U.S. Environmental Protection Agency

the nitrogen cycle (similar to how CO_2 is fixed by plants), its average duration in the atmosphere is 114 years, longer than that of many other GHGs. Unlike some other GHGs, however, N_2O emissions have been relatively constant for the past two decades, making its role in future climatic warming somewhat less central than that of CO_2 and methane.

Chlorofluorocarbons (CFCs) also play a major role in radiative forcing, although their impact has been lessened considerably. Chlorofluorocarbons are a family of gases used primarily in industrial applications and were a primary factor in the depletion of the ozone layer. Several international agreements, most notably the Montreal Protocol, have regulated the production of CFCs, and consequently the emissions of CFCs have dropped significantly. The protocols banning most CFCs have been a particular success in emissions regulations. Unfortunately, many of the substances that have been used as replacements for CFCs, mainly hydrofluorocarbons, are potent GHGs themselves. Hydrofluorocarbon emissions are forecasted to grow almost 150 percent by 2020 compared to their 2005 baseline, and some hydrofluorocarbons are up to 1,200 times more potent per kilogram than CO_2. The replacement of CFCs with other pollutant gases illustrates the difficulty in cleaning many industrial processes—an illustrative example of the challenge in phasing out polluting energy sources with others that may have their own environmental risks.

The Climate as a Total System

The increasing radiative forcing caused by GHGs has been the major focus of climate change concerns for the past several decades, but the incredible complexity of the climate as a system makes it difficult to accurately predict future warming trends. Many factors play important roles in climate change, and understanding each of the many interdependent aspects of the climate system is important, not only in attempting to mitigate a degree of climate change, but also in understanding how humanity will need to cope with the effects of a changing global climate.

Greenhouse gases are the primary anthropogenic element driving climate change, but several natural systems and processes impact the severity of the change, alter natural feedback cycles, or delay certain effects. **Albedo** is an important concept in climatology; the albedo of a surface is a measure of how much energy is absorbed versus how much energy is reflected away (Image 15.9). As gases in the atmosphere absorb solar radiation and emit it as heat (rather than reflecting it back into space), so too do the natural features of the earth, such as snow, trees, or the ocean. Darker surfaces have a lower albedo (meaning that they absorb more energy), while lighter surfaces have a higher albedo. This effect is particularly important in the high northern and southern latitudes, where a warming climate will lead to less snow cover and thus less solar energy reflected away from the planet's surface and back into the atmosphere. The darker colored ground or ocean where there was once snow or ice will absorb and radiate more heat. As with methane from permafrost, melting snow can cause a feedback loop, through which absorbed heat in low-albedo surfaces will melt more snow, continually lowering the albedo of the planet.

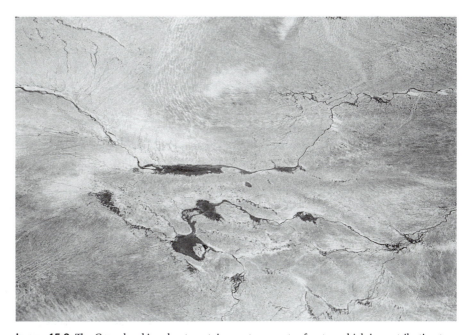

Image 15.8 The Greenland ice sheet contains vast amounts of water, which is contributing to sea level rise as the ice sheet melts.

Milan Petrovic / Shutterstock

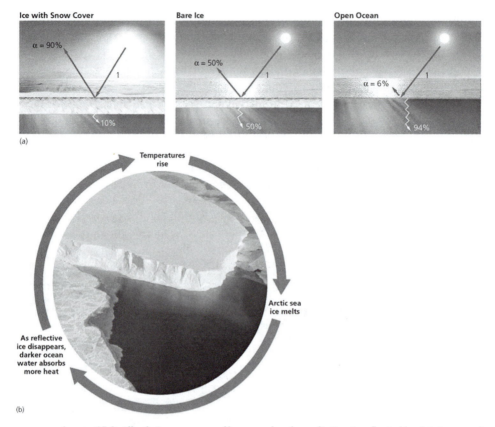

Image 15.9 Albedo is a measure of how much solar radiation is reflected back into space by a surface (α is the percentage of reflected radiation). The darker the surface, the more radiation is absorbed, contributing to further melting and a positive feedback cycle that further decreases the albedo.

One of the most critical factors in understanding and projecting climate changes has to do with the storage of heat in the world's oceans. Because water can absorb such a large amount of heat energy, up to 80 percent of the total excess heat from the past four decades has ended up in the oceans. Even slight temperature changes in oceans can have profound effects on our climate; the El Niño weather pattern is a potent example of how only a few degrees can alter weather worldwide (Image 15.10). Oceanic circulation is strongly affected by small changes in temperature and, similarly, has major effects on weather and climate worldwide. One of the primary difficulties with oceanic warming is that water temperatures can lag decades behind atmospheric conditions because of the massive heat capacity of the world's bodies of water, especially the deep oceans; this makes it difficult to precisely account for the ocean's effect on future climate changes. Water can absorb much more heat energy than most other substances, and much of the heat in the oceans is distributed deeper under water, where it will not have immediate effects

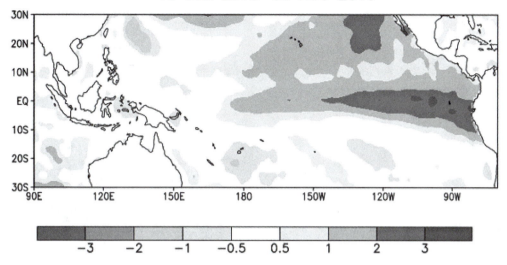

Image 15.10 The El Niño disturbance in the Pacific Ocean impacts weather patterns globally. As the climate changes, we are seeing stronger El Niños than in the past.

on the climate. Much of the stored heat from past decades will start to have a stronger warming effect as oceanic temperatures catch up to atmospheric temperatures, further exacerbating climatic warming.

Oceanic warming is only one of many impacts of climate change on the world's oceans. The worldwide circulation of currents (or **thermohaline circulation**) is being altered as fresh water from melting sea ice and glaciers dilutes the salinity of the oceans, changing the course of ocean currents (Image 15.11). There are numerous potential impacts of oceanic current shifts, from affecting local weather in coastal zones to altering the balance of marine ecosystems worldwide. Warming oceanic temperatures, and a subsequent acidification of the ocean through CO_2 absorption, have also been implicated in the ongoing increases in coral bleaching, which has been damaging or destroying coral reefs in the tropical latitudes (Image 15.12). Warmer temperatures cause an increase in the speed of coral bleaching, which in turn prevents coral reefs from supporting the diverse and plentiful marine life they normally harbor. These declines result in a domino effect, not only harming marine life, but also greatly impacting the health of commercial fish stocks, which can negatively impact the populations that rely on fishing for their livelihoods. Because the oceans are so large and complex, the range of potential climate impacts from warming temperatures is extremely large. From coastal erosion to circulation to fish stocks, the progress of climate change is likely to substantially impact our world's oceans and the people who rely on them.

Image 15.11 The thermohaline circulation helps regulate global temperatures.

Image 15.12 Much of the world's extra carbon dioxide from burning fossil fuels has been absorbed by the ocean, leading to ocean acidification. Coral bleaching is one symptom of ocean acidification.

Ethan Daniels / Shutterstock

Human Emissions from Industrialization to the Present

Gaseous emissions from fossil fuels, followed by land use change, are the primary drivers of anthropogenic climate change. The Industrial Revolution, largely powered by coal, represents the beginning of large-scale human GHG emissions, and global emissions have been increasing ever since (Image 15.13). Industrialization began primarily in Europe and North America, but has since spread to much of the rest of the world. The world's most developed countries, because of their extremely high demand for energy, are still the largest emitters of GHGs per capita, although many regions have different levels of emissions. Environmental regulations in the most highly developed regions, in addition to the faster growth of renewable energy systems in wealthy countries, have led to a decrease in per capita emissions in these regions, but poorer and rapidly developing countries have tended to rely on dirtier (and cheaper) fuel sources such as coal.

The United States is no longer the world's largest emitter of GHGs, having been displaced by China in 2005. Regardless, the United States is still the second largest emitter, responsible for 12.9 percent of all GHGs in 2013, or 6,213 million metric tons of CO_2 equivalent. The European Union (EU), which has a significantly larger population, but also a wide range of varying levels of industrialization and different regulatory systems, comes in as the third largest emitter, responsible for 7.9 percent of global emissions, or 3,795 million metric tons of CO_2 equivalent. Between the United States and the EU, the world's most advanced economies (excluding Japan and other highly developed countries in East Asia) emit more than one-fifth of all the world's GHGs, despite having only 11 percent of the global population. Their per capita emissions align closely with their per capita energy

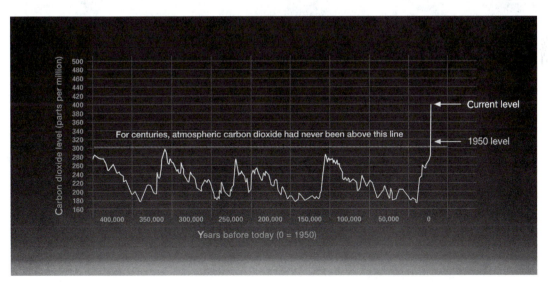

Image 15.13 The concentrations of carbon dioxide in the atmosphere have grown dramatically since the Industrial Revolution.

Courtesy of the National Aeronautics and Space Administration usage, with the United States being the twentieth most prolific emitter (19.9 metric tons of CO_2 equivalent per person per year), while several EU countries are in the top twenty-five. Much of the higher energy use per capita in the United States can be explained by the design of American cities, their transit planning, and their energy infrastructure. The low density of U.S. cities necessitates significantly more personal car usage, while large homes and high consumer spending drive increased energy usage.

The earliest industrialized countries (the United States, the EU, and Japan) are still some of the largest GHG emitters both in general and per capita, but as developing countries further industrialize and use more and more energy, they will catch up and exceed the GHG emission levels of the developed world. China, for its part, has already exceeded the total emissions of the United States, emitting 22.7 percent (9,679.3 million tons of CO_2 equivalent) of GHGs in 2010, with India growing quickly as well, becoming the third largest emitter in recent years. Because China and India have such large populations (37 percent of the global populace between the two) and are continuing to grow, more modest increases in per capita energy usage result in outsized changes in their national totals (Image 15.15). As emissions from some of the larger developing

Image 15.14 Automobiles are a major source of greenhouse gas pollutants and will remain so until we stop using petroleum as our primary transportation fuel.
Chuyuss / Shutterstock

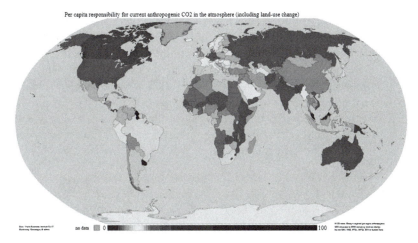

Per capita responsibility for current anthropogenic CO2 in the atmosphere (including land-use change)

Image 15.15 Greenhouse gas emissions are most intense in highly developed countries on a per capita basis. Other areas with intense per capita emissions include countries on the Arabian Peninsula, where oil and gas are highly subsidized.

User:Vinny Burgoo / Wikimedia Commons / CC-BY-SA-3.0

countries increase, many of their governments are beginning to implement mitigation plans and invest in renewable energy technologies.

Impacts of a Changing Climate

There is a wide range of current and potential future impacts from climate change. In addition to the most apparent effect in warming temperatures, many interconnected natural systems will be affected in turn, in a sort of domino effect. For example, melting glaciers will change the temperature and salinity of seawater, which can alter weather patterns and change sea level, causing social disruption and even migration and conflict. As the impacts from certain effects of climate change become more and more complicated, they are more difficult to accurately forecast, but the subtle influence of even a few degrees of extra warmth can have a greatly exaggerated effect on numerous systems as they interact with each other.

The most clearly observable effects of a warming climate at present are taking place in the Arctic and Antarctic regions of the world (Image 15.16). As a result of several physical systems, such as albedo and permafrost melting effects described earlier, warming happens much more quickly in the polar regions, particularly the Arctic, a phenomenon known as **Arctic amplification**. Because of this rapid warming, sea ice in the Arctic has been shrinking at a fast pace, while glacial coverage (particularly in Greenland) has been dropping precipitously. Melting sea ice does not itself represent a major negative impact because it will not affect sea levels, but ice sheets and glaciers on land, especially the largest glaciers in Greenland

and Antarctica, hold massive amounts of fresh water that would cause substantial rises in sea level if they were all to melt. Estimates by NASA indicate that Greenland's ice sheets alone hold about 8 percent of the world's fresh water and would increase sea levels by up to 7 meters (23 feet) if they were completely melted (Image 15.17). Further, the colossal influx of fresh water would change the salinity of the world's oceans, altering oceanic circulation, wind patterns, and weather worldwide. Current models do not predict that all ice sheets will melt within the next hundred years, but at current rates, a certain amount of sea level rise caused by melting is virtually guaranteed.

Image 15.16 The loss of sea ice is impacting many animals in the Arctic.
Outdoorsman / Shutterstock

Image 15.17 Reduced sea ice is opening up shipping, but having many negative consequences for wildlife and Arctic communities.
Denis Burdin / Shutterstock

The anticipated rise in sea levels is a major physical threat to much of the world's population—the United Nations notes that half of the world's population lives within 60 kilometers (37 miles) of a coast, and 75 percent of the world's major cities lie directly on coastlines. A sea level rise of just 1 meter (3 feet) would flood many of the world's major cities, in addition to destroying many natural ecosystems, rendering large amounts of farmland useless and increasing coastal erosion. The most recent **Intergovernmental Panel on Climate Change** (IPCC) report has revealed that sea level has been rising faster than previous climate models suggested and that levels may rise between 28 and 98 centimeters (11 to 39 inches) by the end of the century (Image 15.19). The consequences of such a large rise in sea levels in only a few decades would be serious for coastal and other low-lying communities and could result in many major cities worldwide either being partially submerged or compelled to implement extensive mitigation measures at great expense to stay dry below sea level.

Sea Level Rise

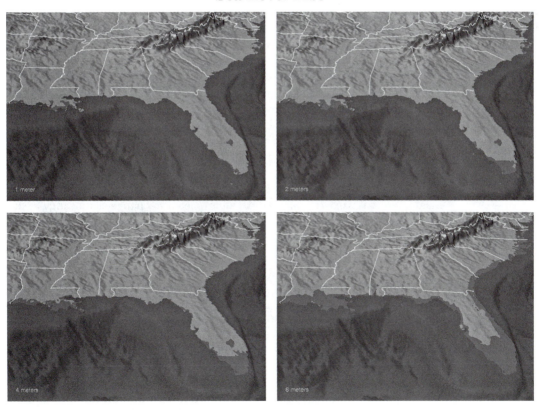

Image 15.18 Sea level rise could lead to the loss of coastal communities and imperil many low-lying areas, including islands and other low areas like Florida.

Courtesy of the National Oceanic and Atmospheric Administration, Geophysical Fluid Dynamics Laboratory

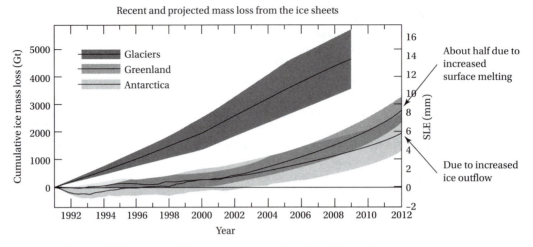

Recent and projected mass loss from the ice sheets

Image 15.19 Courtesy of the Intergovernmental Panel on Climate Change

One of the biggest potential challenges of a warming climate would come from the major shifts in worldwide weather patterns. A great deal of public attention has focused in particular on the strengthening of tropical storms and hurricanes, one of the more immediately destructive effects of warming oceans. The U.S. National Oceanic and Atmospheric Administration estimates that projected changes in temperatures based on the most recent IPCC report could strengthen the intensity of hurricanes and tropical cyclones by up to 11 percent by the end of the twenty-first century, and the incidence of more destructive storms will likely increase by a much greater percentage. Consequently, tropical storms in the next century will be more destructive, which will come at a great cost to human lives and coastal economies (Image 15.20).

In addition to storms, rainfall patterns are expected to change significantly as the progress of warming continues. Changes in climate have already caused perceptible increases in worldwide precipitation levels, and warmer temperatures will continue this pattern. In general, wet places around the world will get wetter, while dry places will get drier; much of the precipitation from midlatitudes will move toward the poles, resulting in wetter temperate regions and drier warm regions. This will exacerbate the impact of droughts on many places where they are already commonplace (particularly in heavily populated regions like West Africa and the American Deep South), and large floods will likely become more common in many regions. As with oceanic changes, it is difficult to more precisely forecast the exact changes in rainfall, but worldwide rainfall models illustrate the expected general trends.

The impacts of a warming climate are diverse and difficult to accurately predict, but the advances in climate science over the past decades have painted an alarming picture of what humanity can expect to encounter in the coming decades. Changes in weather, more intense storms, longer droughts, and worse floods are a few of the many likely impacts from forecasted climate changes. These initial physical impacts are only the first layer of how a warmer climate will truly impact humanity; these changes will bring

Image 15.20 Climate change is leading to stronger storms, such as Superstorm Sandy, which impacted the U.S. East Coast, including New York City. Courtesy of NASA Earth Observatory / Jesse Allen / Science Source.

NASA Earth Observatory/Jesse Allen / Science Source

about a sort of domino effect for human society, altering global agriculture and exacerbating population movements, water resources, and more. Because so many of the impacts could have such dire consequences for our society, we must examine our use of energy resources and their emissions and attempt to limit our ongoing contributions to climate change.

Toward a Solution

Global climate change is complicated, and the vast number of interdependent variables makes its scientific study difficult and imprecise. Although most scientists accept the assertion that climate change is real and manmade, the slow and methodical progress of scientific research, along with the disagreement on some specifics of the science, makes it difficult for scientists to communicate the importance of their work to the general public. Many pundits take any disagreement within the scientific community as a sign that climate change is "merely a theory," mistaking disagreement for dissent. Consequently, imposing rules and regulations to mitigate the effects of climate change is extraordinarily difficult when much of the public cannot even be convinced that the phenomenon is real. Organizations such as the IPCC have been formed in response to public skepticism. These organizations present reports built through consensus of thousands of scientists in all fields relevant to climate change, providing something of a unified front to present the evidence, progress, and probable impacts of the changing climate.

Despite the efforts of scientists worldwide and the overwhelming nature of the scientific evidence in favor of a changing climate, imposing solutions to climate change is difficult. For those who deny the reality of climate change, or at least have yet to concretely feel its current impacts, the economic and political costs of proposed solutions can often seem too high to make action worthwhile. Cutting emissions requires either using less energy or switching to cleaner (and often more expensive) sources of energy. Preserving the environment and sacrificing economic growth are unattractive solutions for many corporations and governments that are intent on growth and development as their highest priorities. Developing countries often raise the objection

Image 15.21 There is growing international pressure, including from the public, to address climate change more forcefully.

Rena Schild / Shutterstock

Image 15.22 Changes to the climate are likely to impact the poorest and most marginal people the most, such as those who rely on subsistence agriculture and fishing to survive.
Gail Johnson / Shutterstock

that industrialized societies could pollute as much as possible to lift their citizens out of poverty in the past and that it is unfair for the international community to hold these countries to a different standard because other countries have caused most of the climatic changes thus far. Because of the great difficulty in convincing both policy makers and the public that climate change is real and must be mitigated, there is an uphill battle before the challenge of implementing policy to combat climate change can be addressed.

One of the biggest challenges in implementing policy to combat climate change can be explained with the idea of the **tragedy of the commons**. The climate is shared by all people and countries on earth without respect to borders, but change must be implemented by individuals. A good analogy for the tragedy of the commons is the challenge of regulating fishing outside of territorial waters. While one fisherman may reduce his catch of a limited supply to preserve stock, others may take advantage of the plentiful resource and catch as many fish as possible, which will lead to declining supplies despite the conservation efforts of the one fisherman. Similarly, some countries may implement rules to reduce emissions, but other countries may assume that their emissions are unimportant (or that they are offset by the conservation efforts of others) and will thus prioritize their own economic growth at the expense of everyone's environment. Getting everyone to cooperate to preserve a communal resource is difficult and requires complete participation. With a diverse community of countries across the world, achieving a consensus on what rules are acceptable is complicated. Until rules are agreed on, most countries act in their own economic interest, resulting in increased emissions and further exacerbation of the changing climate.

Despite the lack of a unified international community, there have been several international efforts to combat rising emissions levels and slow the progress of climate change. Early efforts such as the Montreal Protocol (in 1989) limited the use of ozone-depleting chemicals among signatories, while later treaty efforts such as the Kyoto Protocol (in 1997) aimed to commit signatories to gradually reduce all GHG emissions over time. Treaties with legally binding limits on emissions have been far from universally accepted; in the past, industrializing countries such as China and India have rejected emissions limits because they impede these countries' ability to grow without restrictions, while countries like the United States cannot generally come to a consensus on highly politicized issues like climate change. Further agreements such as the Copenhagen Accord of 2009 encourage countries to limit their emissions, but also lack the weight of a treaty or other international legal agreement and are not binding. The inability to convince most countries to limit to emissions regulations further illustrates the difficulties of the tragedy of the commons and indicates that mitigating climate change will require more than a broad international political framework.

Many countries are becoming more invested in mitigating climate change, even as some major emitters refuse to be strictly bound by international climate treaties. China's rapid industrial development has led to extensive pollution across the country and major health problems for the population. Even without the specter of climate change, China's government has realized the serious costs imposed by its industrialization; consequently, there has been a major drive to implement renewable energy technologies to provide a cleaner environment for the population. Many other developed countries have taken advantage of dropping wind and solar prices to begin a shift toward renewable energy in their grids, both to come in line with climate change goals and to improve the quality of life for their citizens. Many of the current international policy discussions on climate change are taking these energy switches into account for setting future emissions goals, and a combination of policy changes, technology, and lifestyle changes may be able to keep future warming below the 2°C mark that is often set as the standard for the maximum amount we can allow the climate to warm before the most disastrous complications set in.

In late 2015, the most significant climate agreement since the Kyoto Protocol was adopted at the Paris meeting of the Conference of Parties of the United Nations Framework Convention on Climate Change (Image 15.23). The new Paris Agreement is the first climate agreement with almost universal agreement that set limits on all countries, including developing countries, regarding GHG emissions in nationally determined contributions. In addition, the new agreement continues policies laid out in the IPCC that call on developed countries to provide assistance to poorer countries to help with climate change adaption and mitigation, including deploying more clean energy. The agreement also calls for countries to update their plans every five years. How this is implemented, or not implemented, will significantly impact the overall approach to the climate.

At their core, climate change and global warming are directly proportional to the energy use of humanity. The primary driver of climate change is the emission of GHGs, the majority of which are emitted during the combustion of fossil fuels. A certain degree of climate change is already guaranteed and, indeed, has already come to pass because

Image 15.23 The Paris Conference of the Parties agreement is the most thorough climate change agreement yet reached.
WITT/SIPA / Associated Press

of the vast amounts of GHGs that have been released into the atmosphere over the past century or so, but humanity still has the chance to curtail some of the growth in emissions and avoid the worst-case scenarios of global warming. To head off many catastrophic consequences of climate change, however, the energy choices we make as a society must be thoroughly examined.

Renewable energy technologies represent the greatest hope for greatly reducing GHG emissions and slowing the onset of warming, yet many major obstacles stand in the way of a transition to a completely renewable energy system. As discussed in their respective chapters, wind and solar both suffer from variability; without the development of a large-scale grid storage system, there is no way to ensure that a grid remains powered when the sun is not shining and the wind is not blowing. A diverse mix of energy generation types can ameliorate some of this variability, but most grids cannot be feasibly powered entirely by renewable sources yet. Technological advances in electricity generation, improvements in smart grids, a broader variety of renewable technologies, and backup generation from sources like natural gas, nuclear, and hydro can help increase the share of emissions-free energy and slow the contribution of emissions to anthropogenic climate change, but technology alone is unlikely to be sufficient. Global and local political frameworks, emissions reductions, and united efforts to promote more sustainable levels of energy usage will be required if we are to avoid some of the worst possible scenarios of climate change as outlined in scientific reviews such as the IPCC report. For further information on the most current scientific research and forecasts related to climate change, see the IPCC reports at www.ipcc.ch.

REVIEW QUESTIONS

1. Why is climate change inherently an energy issue?
2. Within the scientific community, there is continued debate related to climate change, but not its existence. Where is there uncertainty, and why?
3. What role will the ocean play in future climate change? How about reductions in ice and snow cover?

Image 16.1 Associated Press

CHAPTER 16

The Geopolitical Challenges of Energy

Instability in the Middle East, covert nuclear programs in North Korea and Iran, Russian gas supplies to Europe cut off—these are but some examples of the nexus between the global energy system and international politics. Although we trade all of the various energy resources through the global economy, only a few have major impacts on the geopolitical situation. Because large amounts of coal remain in the control of the Western developed countries, especially the United States and Australia, there is far less instability associated with this resource. However, both oil and natural gas are impacted by major geopolitical issues. Because of the dual-use nature of nuclear power providing the potential for the development of the most dangerous weapons humankind has ever made, there is inherent instability associated with the spread of nuclear technology to additional countries as well. These instabilities are not simply tied to nonrenewable resources, because one renewable, hydropower, also can have geopolitical implications, such as when upstream states on rivers that cross borders construct dams that restrict the flow of much-needed water to downstream states.

We will explore the nexus between global politics and energy in this chapter, focusing on the four resources most implicated in geopolitical instability—oil, natural gas, nuclear, and hydropower—looking at the players in the international system and the specific challenges associated with each of these resources.

Oil Geopolitics

Oil is one of, if not the most, volatile commodity traded on the global market. As the backbone of the modern globalized economy, the free flow of oil at a relatively low price remains essential for global economic health. However, much of the world's oil comes from politically unstable places, including countries that are isolated in the global system. Oil has made the Middle East incredibly wealthy, but has allowed this vast wealth to fall into the hands of corrupt autocratic governments and terrorist organizations. Other oil exporters have also faced major political instability at home and exported such instability abroad (Image 16.2).

Although large amounts of oil continue to be produced in consumer countries like the United States and China, most global oil exports come from a variety of emerging countries. The Middle East and North Africa region dominates the major oil exporters list, with three of the top five exporters—Saudi Arabia, the United Arab Emirates, and Iraq (Table 16.1). The same can be said for proven reserves of oil, with three of the top five and six of the top ten coming from the Middle East and North Africa region (*Note*: Number one Venezuela and number three Canada owe their positions to their vast deposits of oil sands, while the Middle Eastern countries sit on vast reserves of liquid petroleum, which is far easier to extract, refine, and bring to the global marketplace) (Table 16.2).

Image 16.2 Collapsing prices for oil helped lead to growing instability in petro-states like Venezuela.

Fernando Llano / Associated Press

Table 16.1 Top Five Global Oil Exporters (barrels per day)

Saudi Arabia	7,416,000
Russia	4,888,000
Iraq	3,301,000
Canada	3,210,000
United Arab Emirates	2,637,000

Source: CIA World Factbook. https://www.cia.gov/library/publications/the-world-factbook/rankorder/2242rank.html

Table 16.2 Top Five Global Oil Proved Reserves (billion barrels)

Venezuela	300.0
Saudi Arabia	269.0
Canada	171.0
Iran	157.8
Iraq	140.0

Source: CIA World Factbook. https://www.cia.gov/library/publications/the-world-factbook/rankorder/2244rank.html

The preponderance of unstable states among global oil exporters and holders of oil reserves is one of the major causes of the instability of global oil prices (Image 16.3). Since the year 2000, oil prices have been as low as $25 per barrel and as high as $147 per barrel. Such volatility adds enormous uncertainty to the global economy because of oil's predominant role in the global transportation system and as a feedstock for various products that are central to the economy. Price spikes in oil have been caused by wars, such as the 1990 spike when Iraq invaded Kuwait, and are also influenced by other forms of instability, especially in the Middle East, Nigeria, and Venezuela.

There are strong demonstrated links between oil exports and instability in developing countries. Oil-producing states sit at the center of many of today's major conflicts and areas of civil unrest, from Iraq to Nigeria to Venezuela (Image 16.4). One of the driving factors of this instability is **Dutch disease**, named after the economic challenges suffered in the Netherlands after major oil and gas discoveries there in the 1960s. This problem is caused by rising resource exports, which drive up currency prices, which in turn makes it cheaper to import other products, rather than producing them. This cycle causes a loss in economic diversity, leading to enormous problems when the price falls.

Additionally, these huge oil windfalls often go on to line the pockets of the rich and powerful, rather than being reinvested in the country at large. An example of this is Equatorial Guinea, which has the largest gap between per capita gross domestic product and human development in the world. On paper, Equatorial Guinea is the wealthiest country in Africa when measured per capita at over $32,000 per person; however, 77 percent of the population lives below the poverty line and most of the wealth is

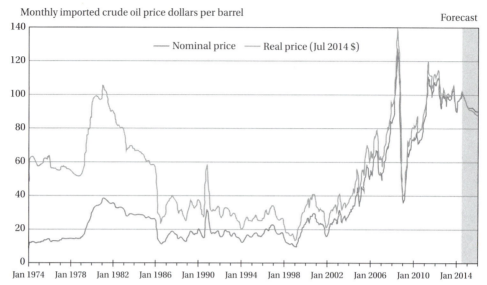

Monthly imported crude oil price dollars per barrel

Forecast

Image 16.3 Oil prices are volatile. Geopolitical issues weigh heavily on these changes.
Courtesy of the U.S. Energy Information Administration

Image 16.4 Oil development in Nigeria has enriched the few individuals with political connections, while not providing additional income to local residents in the oil-producing areas.
Navin Mistry / Shutterstock

concentrated in the hands of the president for life, Teodoro Obiang, and his extended family (Image 16.5).

Oil is unique among energy resources in that large portions of the global oil supply are controlled by a **cartel**—in this case the **Organization of the Petroleum Exporting Countries** (**OPEC**) (Image 16.6). The organization is not able to fully control the global price of oil, but is able to help shape it because it sets quotas on monthly production to keep oil prices within a specific range. Members of the organization produce about 40 percent of global oil and 60 percent of that which is traded internationally. Beyond that, OPEC countries hold about two-thirds of global oil reserves. Saudi Arabia is OPEC's principal producer, accounting for 13 percent of global oil production in 2013 and over 30 percent of OPEC production. More than any other OPEC member, Saudi Arabia is able to adjust its oil production to help meet OPEC quotas, making it the global *swing producer* of crude oil. Despite Saudi Arabia's role, other OPEC members do not always go along. Often, they continue to produce at the same rates, even when OPEC agrees to a production cut, leading Saudi Arabia to bear the brunt of change. When oil prices have spiked, Saudi Arabia's swing capacity has not always been enough to bring oil prices down to the level that OPEC is targeting, as seen in the 2008 price spike. Conversely, when oil prices have slumped, Saudi Arabia has not always been willing to cut back on production to reverse the price slide, as happened in 2015.

The Organization of the Petroleum Exporting Countries rose to prominence only a decade after its founding, during the Arab–Israeli conflicts of the early 1970s. In 1973, in response to the Yom Kippur War in which the United States sided with Israel, OPEC announced an embargo against the United States and its allies (Image 16.7). This led to a fourfold increase in world oil prices and clearly placed OPEC on the map as a major player in the geopolitical system.

Image 16.5 No country better represents the imbalance of oil revenues than Equatorial Guinea, where President Obiang (pictured) and his family control most of the economy. Despite having the highest per capita gross domestic product in Africa, most residents live on one to two dollars per day.
John Minchillo / Associated Press

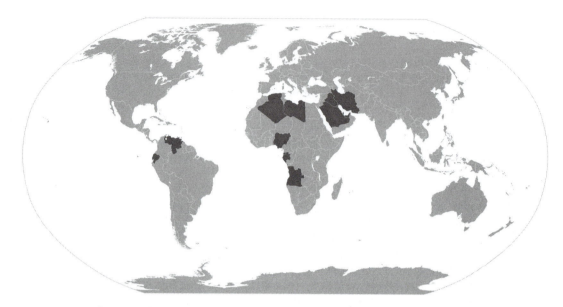

Image 16.6 The Organization of the Petroleum Exporting Countries remains a small organization, with most of its members coming from the Middle East and North Africa.
User:Bourgeois / Wikimedia Commons / CC-BY-SA-3.0

Image 16.7 The 1973 oil crisis, when the Organization of the Petroleum Exporting Countries cut off exports to countries supportive of Israel, led to major shortages of gasoline in the United States.
Associated Press

Today, OPEC consists of thirteen states, with eight located in the volatile Middle East and North Africa region (Image 16.8). Two major global players, Venezuela and Nigeria, fall outside this realm, along with three smaller OPEC players, Ecuador, Indonesia, and Angola. Venezuela and Nigeria are also highly unstable countries suffering from severe

Image 16.8 The Organization of the Petroleum Exporting Countries meets regularly in Vienna, where it is headquartered, to discuss changes to global oil output by its member states.
Kyodo News / Associated Press

governance problems. Venezuela made itself a major adversary of the United States under its previous president, Hugo Chávez, and the situation continued under President Nicolás Maduro. Nigeria, although friendly to the United States and the West more generally, is beset with some of the world's worst corruption, a major terrorist insurgency led by Boko Haram, and decades-old major unrest in the Niger Delta, where most of the oil deposits are located. Similarly, many of the Middle Eastern OPEC members face severe challenges that impact their ability to be reliable producers of oil. Iraq and Libya are beset by internal strife and terrorism, while Iran's oil industry has been impacted by international sanctions related to their nuclear program (see more in the section on nuclear proliferation in this chapter). The other OPEC members in the Middle East are currently stable, but most are governed by monarchies and are susceptible to the broader unrest that has best Libya, Iraq, Syria, and Yemen, among others, especially since the 2003 Iraq invasion and the 2011 Arab Spring. In short, we can expect continued challenges in OPEC countries and their neighbors that will continue to impact their ability to supply the global market in a predictable and stable manner.

Unlike the instability that has plagued so many OPEC producers, the other major players in the global oil export market are generally more stable. Lead non-OPEC producers include Russia, Canada, Kazakhstan, Mexico, and Norway. While Russia has been a destabilizing international actor (see the following section on pipelines and natural gas), the other non-OPEC major exporters are relatively stable and, other than Kazakhstan, are free and democratic countries. This provides a safety valve for the United States, Europe, and Japan, the major developed oil importers (Image 16.9). Despite this, OPEC still dominates imports, with 45 percent of oil imported into the United States coming from OPEC and even larger shares of total imports into the European Union (EU) and Japan.

Another area of growing instability related to oil deposits is the South China Sea, especially the region around the contested Spratly Islands (Image 16.10). Although oil is not

Sources of U.S. net petroleum imports, 2013

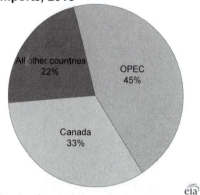

All other countries
22%

OPEC
45%

Canada
33%

Note: Petroleum includes crude oil, petroleum products, and biofuels.

Image 16.9 Combined, the member countries of the Organization of the Petroleum Exporting Countries are the largest source of U.S. oil imports, but Canada is the single largest outside supplier.

Courtesy of the U.S. Energy Information Administration

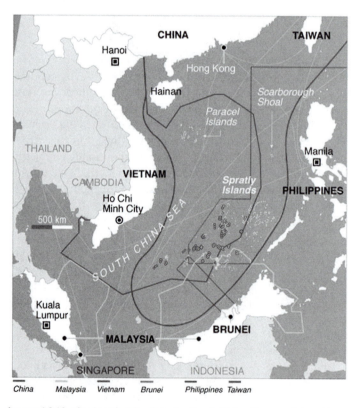

Image 16.10 The Spratly Islands in the South China Sea are claimed by many countries. The possibility of oil deposits in the area is one of the main reasons so many states want sovereignty.

Voice of America / Wikimedia Commons / CC-BY-SA-3.0

Image 16.11 The Philippines, China, and Vietnam are building bases in the Spratlys to further cement their claims to the islands and their surrounding waters.
Kyodo News / Associated Press

produced in this area now, it is surmised that it contains large oil deposits. The Spratlys are a challenge because they are claimed by several countries that surround the South China Sea—China, Vietnam, Malaysia, Brunei, Taiwan, and the Philippines—in whole or in part. They are also uninhabited small reefs that barely emerge from the water. Recently, China has embarked on a major campaign to build military bases on these islands by dredging sand and trying to build up the islands. Other countries that make claims have engaged in similar actions and there are now regular naval incidents in the area, some that even involve the U.S. Navy (Image 16.11). Although no oil is being produced now, the presence of potential oil helps drive this territorial dispute, one that is likely to grow in the coming decades.

Despite surging oil production in the United States, large portions of the oil needed in the United States and throughout the other major energy-consuming regions will have to come from unstable places. Most projections for future energy consumption assume massive growth in oil consumption globally, with China and India emerging as the largest new consumers. These same projections assume little to no reduction in overall consumption by the developed countries. As such, more players will be chasing after the same suppliers, leading to a growth in OPEC's role in the global oil economy because they hold the majority of oil reserves.

Pipelines and Natural Gas Geopolitics

Although the **geopolitics** of oil have been a fixture in international politics since the 1970s oil crises and earlier, the politics of natural gas on the international scene are a more recent phenomenon. Most of the major issues surrounding natural gas in the geopolitical system revolve around the routes of pipelines that cross international borders, especially pipelines from Russia to European consumer countries. Because of the large additional costs and limited supply of liquefied natural gas (LNG), these pipelines and their routes have played a major role in international disputes, especially during the early twenty-first century.

Unlike oil, which is easily transported via ship and commands a relatively consistent global price, natural gas requires significant energy and cost to liquefy, leading to vast differences in gas prices from region to region. Geopolitical factors have had enormous impacts on price, especially in the Eurasian context where Russia is the dominant exporter (Table 16.3), while EU countries are the primary importers. In fact, Russia is by far the largest global natural gas exporter, with around 200 billion cubic meters in exports, while the EU, as a bloc, is the world's largest importer, with over 420 billion cubic meters in total imports. Although the EU imports natural gas from Norway, Algeria, and the Caucasus countries via pipeline through Turkey, Russia is the largest source, and it has exploited its position in significant ways.

Russia has used low gas prices to try to lure former Soviet neighbors, such as Belarus and, until recently, Ukraine, to keep them in its orbit. However, most of the pipelines that serve higher paying customers to the west, especially Germany, cross Ukraine (Image 16.12). As such, gas disputes over pricing between Ukraine and Russia have led to a series of gas crises in Europe. Each time there has been a major dispute, Russia has cut off or dramatically curtailed gas supplies to Ukraine, leading to supply disruptions in the EU and often to Ukraine siphoning off the remaining natural gas for their own use. Following the unrest and outbreak of war in 2014–15, the situation has become even more precarious.

Because of these disruptions, Russia has proposed two new pipelines—North Stream and South Stream—which pass through the Baltic and Black Seas, respectively, as a means to bypass the large number of international borders that their existing pipelines cover. Due to Russia's isolation and economic problems, following interventions in Ukraine and Syria, there is currently no funding for these projects. In addition, these disputes, as well as general European mistrust of Russia's intentions, have led to alternative suppliers stepping in (Image 16.13). Gas from Azerbaijan now passes to the EU via Georgia and Turkey, while new undersea pipelines connect Algeria with Europe via the Mediterranean Sea.

The lack of stable routes and limited suppliers have given a few countries significant leverage in the natural gas market. Another recent change has been the vast increase in LNG supplies to the global market (Image 16.14). Liquefied natural gas commands a far higher price, which means that markets only choose to purchase it as a last resort to

Table 16.3 Top Five Global Natural Gas–Exporting Countries (billion cubic meters)

Russia	184.5
Qatar	118.9
Norway	114.4
Canada	77.96
Netherlands	53.65

Source: CIA World Factbook. Available at, https://www.cia.gov/library/publications/the-world-factbook/rankorder/2251rank.html

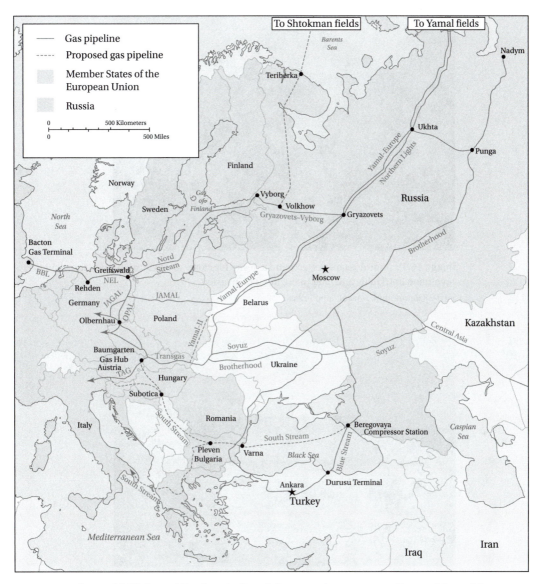

Image 16.12 Several Russian pipelines bring natural gas to western Europe. Many of these pass through Ukraine.

pipelined supplies of natural gas. Thus, the largest importer of LNG is Japan, which has no access to pipelined gas and has needed far more since the Fukushima nuclear disaster led to the closure of most nuclear power plants. China is another major LNG consumer because Russia cannot provide enough natural gas to meet both Chinese and EU demand.

New global energy powerhouses have been launched by these changes, foremost among them Qatar. The Persian Gulf region is disconnected from the broader Eurasian natural gas

Image 16.13 Anti-Russian protests and a desire for closer ties to Europe led to the Maidan revolution in Ukraine and later to the Russian invasion.

Pavlo Palamarchuk / Associated Press

Image 16.14 Because of concerns over Russian natural gas, more liquefied natural gas is coming to Europe from the Middle East and other suppliers.

Aleksander Kamasi / Shutterstock

pipeline network. Because of ongoing geopolitical challenges, especially in Iraq and Syria, Qatar and the other Gulf states (Saudi Arabia, Kuwait, and the United Arab Emirates) are disconnected from the Eurasian network of pipelines. They must therefore liquefy their exports, which primarily go to East Asia. Qatar is now the world's number two natural gas exporter and number one LNG exporter, which, because of fast rising demand, has led them to have the highest per capita gross domestic product in the world. Despite its small size, Qatar has become a geopolitically significant player, all because of LNG (Image 16.15).

Qatar faces significant challenges geopolitically related to natural gas for another border-related reason, that its primary natural gas deposit crosses its maritime boundary with Iran. Qatar's North Field is the world's largest natural gas deposit ever discovered, with an estimated 51 trillion cubic meters of natural gas (Image 16.16). However,

Image 16.15 The skyline of Doha, Qatar, and its bustling economy are being powered by Qatar's major role as a liquefied natural gas exporter.

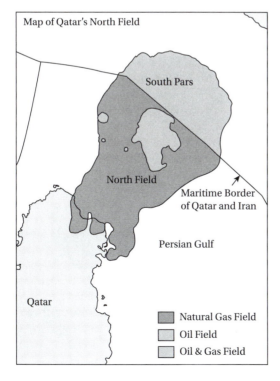

Image 16.16 The North Field in Qatar (called South Pars in Iran) is the world's largest natural gas field.

US Energy Information Administration / Wikimedia Commons / CC-BY-SA-3.0

part of this field crosses into Iranian waters and Iran claims that it is entitled to revenue from the gas field. This alone helps ensure that Qatar remains closely in the orbit of the United States and Saudi Arabia, two countries with long-standing conflicts with Iran. Qatar has gone forward with developing the resource, however, and has been able to leverage the enormous windfall from rising global LNG demand to fund massive public works projects, including the 2022 FIFA World Cup.

Just as LNG has launched Qatar's prominent global role, other major LNG producers hope to do the same. Malaysia, Indonesia, Australia, Nigeria, and Trinidad and Tobago have emerged as other major LNG producers, although Qatar dwarfs them all. The United States, because of the boom in hydraulic fracturing, may soon join the market as well, vying with Russia for a dominant place in the global natural gas market. This newfound competition has become especially significant in the Eastern Hemisphere, where natural gas politics are incredibly complex.

Nuclear Proliferation

Although oil and gas dominate many international discussions of the intersection of energy and politics, no single issue may be more fraught than the link between global politics and nuclear power. Although the risk of a meltdown is one of the primary reasons for objection to nuclear power, its relation to nuclear weapons proliferation is the driving geopolitical concern because raw nuclear materials used for power generation are the same source materials as those for nuclear weapons (Image 16.17). In particular, one of the byproducts of nuclear fission, plutonium, is the other key ingredient that can be used besides uranium to build a nuclear bomb.

Image 16.17 Nuclear fuel processing can provide material for both weapons and nuclear power plants.
Vahid Salemi / Associated Press

The raw materials for nuclear weapons must be enriched to a far higher grade than those used for nuclear power. Uranium for fission reactors is usually enriched to somewhere between 3 and 10 percent U-235 (naturally occurring uranium is about 0.7 percent U-235, with U-238 making up over 99 percent). For nuclear weapons, one must produce **highly enriched uranium**, which is usually around 90 percent U-235. Despite this massive increase, the same technologies are needed; thus, there remains great concern over how many countries, and which ones, have the technology to enrich uranium for reactor fuel. Other nuclear weapons use plutonium, which is a byproduct of nuclear fission in a nuclear reactor.

Several basic types of nuclear weapons can be produced. The two simplest types use nuclear fission reactions to cause an uncontrolled chain reaction (Image 16.18). The most simple, the so-called gun-type weapon, uses a conventional explosive to smash two pieces of U-235 together to cause a fission reaction. The second type uses focused explosions to compress a core of plutonium together in a so-called implosion weapon. More advanced nuclear weapons designs, which cause a far larger explosion, are thermonuclear weapons, including the hydrogen bomb. These utilize an uncontrolled and split-second fusion reaction, but are much more complex to design and build than conventional fission weapons. In any case, these weapons need the

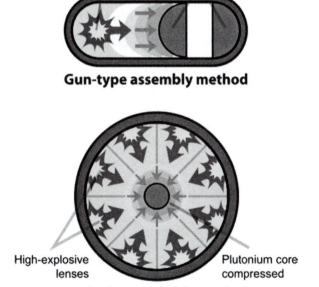

Image 16.18 Nuclear fission weapons use either highly enriched uranium or plutonium (produced as a byproduct of nuclear fission in a reactor) to fuel the warhead.

Fastfission / Wikimedia Commons / CC-BY-SA-3.0

development of a sophisticated nuclear program and the production of appropriate nuclear materials to succeed.

The first nuclear weapons were developed by the United States under the secret **Manhattan Project** program during World War II. Nazi Germany had also been pursuing nuclear weapons, but did not successfully develop their program. The United States first tested a nuclear weapon on July 16, 1945, at the Trinity site in New Mexico. The only two nuclear weapons ever used in war were then deployed on August 6, 1945, at Hiroshima, Japan (a gun-type bomb; Image 16.19), and August 9, 1945, at Nagasaki, Japan (a plutonium implosion bomb). Together, these attacks resulted in more than two hundred thousand deaths.

After World War II, the Soviet Union rushed to catch up and tested their first nuclear weapon on August 29, 1949. The United Kingdom followed in 1952, France in 1960, and China in 1964. These tests are an important marker because a country is not recognized as a nuclear weapons state without first proving their program works by testing a weapon.

Globally, nuclear programs are regulated through the **Nuclear Non-proliferation Treaty** (NPT) and its accompanying United Nations (UN) entity, the **International Atomic Energy Agency**, thus showing the close link between nuclear power and nuclear weapons. These legal regimes aim to control the uses of nuclear technology to

Image 16.19 The first use of a nuclear weapon on a city was the bombing of Hiroshima, Japan, in 1945. A second attack followed days later on Nagasaki, Japan. No other city has ever been the target of a nuclear attack.

Associated Press

ensure its peaceful use, as first outlined in President Dwight Eisenhower's address to the UN General Assembly on Atoms for Peace. In 1960, the NPT was negotiated as the basic framework for the peaceful use of nuclear power without the development of weapons programs, and it came into force in 1968. The NPT guarantees all signatories the right to nuclear power, but with the proviso that no new country develops nuclear weapons. Only five countries were recognized as having the right to nuclear weapons under the NPT: the United States, the Soviet Union (Russia inherited this right in 1991), the United Kingdom, France, and China (Image 16.20). All other countries are subject to inspection of their nuclear facilities to ensure that they are used solely for peaceful purposes.

The NPT quickly became one of the most universally subscribed-to treaties, with almost every UN member signing on. Only four have never signed—Israel, India, Pakistan, and South Sudan—that last of which will likely sign as well. One additional country, North Korea, withdrew from the NPT in 2003.

The International Atomic Energy Agency, created in 1957 as part of the Atoms for Peace program, was formed as the international organization that would provide both technical assistance to help countries develop peaceful nuclear technology and the inspection regimes that would ensure transparency so that countries could not develop nuclear weapons programs in secret. The agency has been embroiled in many nuclear controversies over the years because they provided the UN inspectors that investigated nuclear programs in Iraq, Iran, and Libya.

Since the NPT came into force in 1968, additional countries have defied international law and developed nuclear weapons illegally, thus proving that the danger of

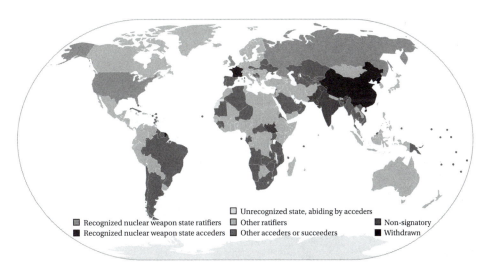

Image 16.20 Most countries are signatory to the Nuclear Non-proliferation Treaty. India, Pakistan, and Israel never signed, and North Korea withdrew.

nuclear proliferation remains a key concern. India joined the nuclear weapons club in 1974, Pakistan in 1998, and North Korea in 2003. South Africa built nuclear weapons during the apartheid era, but then dismantled their program and weapons following the end of apartheid. Argentina and Brazil may have tried to develop weapons as well. Although the country has never tested a weapon, it is believed that Israel has a significant stockpile of plutonium-based nuclear weapons.

Another major concern related to nuclear weapons is derived from their testing. Early nuclear weapons tests were conducted above ground. This led to nuclear radiation fallout far from the test sites, exposing people to harmful doses of radiation. In 1954, the Castle Bravo test of a hydrogen bomb by the United States in the Marshall Islands resulted in an explosion almost twice as large as predicted, which caused massive fallout in areas not predicted to be affected (Image 16.21). Later, the Soviet Union tested the largest nuclear weapon ever made, the Tsar Bomba, over the island of Novaya Zemlya in the Arctic Ocean, resulting in massive destruction and fallout as well. Public awareness of these incidents led to the **Partial Test-Ban Treaty** in 1963, which banned above-ground nuclear testing. In 1996, the **Comprehensive Nuclear Test-Ban Treaty** was negotiated, but it remains not in effect because many countries, including the United States, have not ratified the treaty. Despite not ratifying the treaty, the United States has not tested a nuclear weapon since 1992, and Russia has not tested a weapon since the Soviet Union's collapse.

Image 16.21 In the early years of nuclear weapons, many were tested. This test, at the Bikini Atoll in the Marshall Islands, was an early U.S. nuclear weapons test.
Everett Historical / Shutterstock

The spread of nuclear weapons to more countries provides strong evidence that nuclear power cannot be expanded without great risks. Each additional country with these weapons leads to a more complex geopolitical environment in which unpredictable actors, and perhaps even terrorist groups, could acquire and use nuclear weapons after seventy years of no nuclear weapon attacks. The sheer scale of a nuclear attack makes this especially worrisome. Most recently, the threat of Iran developing a nuclear weapon has led to a massive sanctions regime and nuclear talks between Iran and the permanent members of the UN Security Council, all nuclear weapons states themselves, along with Germany (Image 16.22). In 2015, this led to the signing of the Joint Comprehensive Plan of Action that should result in the freezing of Iran's suspected nuclear weapons program. This agreement could prove to be an important catalyst toward further restrictions on nuclear weapons.

Other major concerns remain because international inspections cannot necessarily halt nuclear programs for countries outside the NPT framework. India and Pakistan now have hundreds of nuclear warheads aimed at each other, and they continue to build more. India has also been able to access peaceful nuclear materials and know-how, including from the United States, without renouncing their program. Pakistan's program is of great concern because their weapons are poorly secured and the country faces tremendous internal instability (Image 16.23). Of even greater worry may be the program in North Korea, arguably the most isolated country in the world. Although their weapons' range is short, North Korea could easily strike Seoul, South Korea, and kill millions before the United States and its allies could respond.

The instability of the nuclear programs in these illegal states has kept the major nuclear powers from agreeing to abolish their own programs, which are primarily

Image 16.22 The 2015 Iranian nuclear agreement was a watershed moment in working to halt nuclear proliferation, while allowing for the use of nuclear power.

Seth Wenig / Associated Press

Image 16.23 The instability of countries like Pakistan, coupled with nuclear arsenals, presents the chance that weapons could fall into the hands of terrorists.
Anjum Naveed / Associated Press

maintained for **nuclear deterrence**. This strategy threatens the use of nuclear weapons solely as retaliation if another country uses them. In effect, a major nuclear power like the United States or Russia could wipe out an entire country if it was to use nuclear weapons.

The United States and Russia have engaged in a dramatic reduction in total nuclear weapons, however (Image 16.24). When the Soviet Union collapsed, three new republics—Ukraine, Belarus, and Kazakhstan—found weapons on their territory, all of which were sent to Russia. They joined with South Africa to form a small group of countries that have relinquished control over the most devastating weapons humankind has ever made. Likewise, the United States and Russia engaged in several rounds of nuclear weapons dismantling, both during and after the Cold War. This has reduced the stockpiles of both countries by thousands of warheads. Further progress will be needed to assuage nuclear power critics to press forward on the peaceful use of nuclear technology.

Water Wars?

Discussions of the relationship between energy and geopolitics often examine fossil fuels and nuclear issues, but alternative energy is rarely discussed. For the most part, this makes sense because technologies such as wind and solar are harvested locally. However, for the most proven renewable, hydropower, there can be important geopolitical connotations when a river basin crosses an international border.

Although no major modern war has centered on conflict over water, there remains great potential in an era of increasing populations and shifting weather patterns.

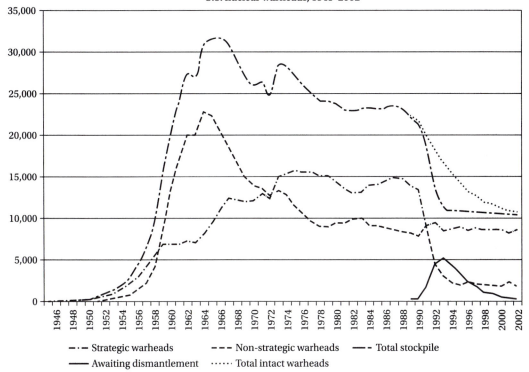

U.S. nuclear warheads, 1945–2002

-·- Strategic warheads   - - - Non-strategic warheads   - - Total stockpile
—— Awaiting dismantlement   ······ Total intact warheads

Image 16.24 A series of agreements with Russia has led to a massive decrease in American and Russian nuclear stockpiles.
Courtesy of the U.S. Department of Defense

Many places in the world are becoming drier but likewise face growing demands for energy. This is especially important to the energy world when hydroelectric dams have been constructed on waterways that cross borders. This construction leads to the creation of reservoirs that hold back water that would otherwise flow to downstream countries. When those countries rely on that water for their own needs, both for the water itself and for energy generation of their own, the recipe for conflict becomes clear.

Most of the world's contested rivers where water is scarce lie in areas that face severe challenges related to population and climate change, such as the Middle East and Africa. Rivers like the Nile, Tigris/Euphrates, and Niger are examples where future water conflict is most likely to occur. Dams are being constructed on these rivers and many already exist (Image 16.25). Without major international efforts to allocate these waters in a fair manner that can be responsive to drought years, as well as years of higher rainfall, the risks of conflict grow.

Future research is likely to focus more on the relationship between water and energy, especially as climate change leads to increasingly dry periods in areas that rely on rivers for drinking water, agriculture, and the generation of electricity.

Image 16.25 Construction of new dams along the Blue Nile in Ethiopia is exacerbating conflict with downstream states, especially Egypt.

Elias Asmare / Associated Press

REVIEW QUESTIONS

1. Why are there fewer geopolitical concerns around coal, wind, or solar power?
2. How is political unrest related to oil prices?
3. Over time, the influence of OPEC has gone up and down. Is it an important actor today and what behaviors among members hurt its influence?
4. Why do pipelines still impact natural gas politics with the development of LNG? Who benefits most from Russian-caused gas crises in Europe?
5. Why are water wars a potential future issue related to hydropower?

Image 17.1 Jaroslava V / Shutterstock

CHAPTER 17

Our Energy Future

After investigating the major energy resources and cross-cutting challenges, predicting the future remains difficult. We are regularly bombarded by news pieces claiming that one resource or plan will somehow dramatically alter our energy system in the next five, ten, or twenty years. Don't count on it! Our energy system will change and in dramatic ways, and these changes will be instigated by changing resource prices, laws and policies regarding pollutants (including carbon dioxide), and technological breakthroughs—none of these can be predicted with certainty. However, two overarching concepts will likely help underpin changes to the energy system: sustainability and **energy switching**. We will examine this as we take a look back at the various resources and concepts covered throughout this book.

Sustainability

Few concepts are as heated academically as the definition of **sustainability** and how it applies to the global energy system. For our purposes, we will rely on the definition first presented at the United Nations with the 1987 Brundtland Report, which defined it as the ability to "meet the needs of the present without compromising the ability of future generations to meet their own needs." The current global energy systems fail to meet this metric. Massive consumption of nonrenewable fossil fuels not only removes these resources from potential use by later generations, but also compromises many of the key planetary systems we need to survive and thrive, including our climate.

But what makes an energy system sustainable? This is where the debate tends to devolve and personal preferences shine through, whether in favor of carbon sequestration, nuclear development, or a total change to renewable resources and abandonment of fossil fuels. None of these is likely, by itself, to be implemented in a manner that lives up to the definition. One could ask whether, barring a major technological breakthrough such as fusion, it is possible to create a sustainable energy system. The answer lies in how one parses this out. On a smaller scale and through changes to one part of the system at a time, we can likely move to a more sustainable energy system. Some of these changes are small, such as the move from incandescent light bulbs to compact fluorescent and light-emitting diode (LED) lights (Image 17.2) or moving more commuters from personal cars to mass-transit options or bicycles (Image 17.3).

More broadly, major changes must be implemented in some sectors of the economy to make the energy system more sustainable, including the following:

- Reorganization of our cities: Denser cities where people live in smaller, more energy-efficient homes and commute without personal cars can dramatically curtail the centrality of the personal car to daily life for many people and reduce individual energy consumption (Image 17.4).
- Green building: More energy-efficient homes, offices, and factories can use less energy for heating, cooling, and light, thus increasing the efficiency with which we go about our daily life and produce the products that people desire and demand (Image 17.5).

Image 17.2 Light-emitting diodes have revolutionized the lighting industry and dramatically reduced the amount of electricity needed for lighting.
Demarco Media / Shutterstock

Image 17.3 Switching from driving cars to riding bicycles reduces congestion, pollution, and energy consumption.
Bikeworldtravel / Shutterstock

Image 17.4 Constructing mixed-use developments where people can live, shop, and work within walking distance is one example of how we can reorganize our cities more sustainably.
Arina Habich / Shutterstock

- Sustainable agriculture: Growing food with lower amounts of water and fertilizer and growing that food locally, when possible, can reduce the energy footprint of both raising food and transporting it (Image 17.6). Making this change fully will require people to understand better where their food comes from and limit the amount of fish, fruit, vegetables, and flowers that are flown long distances to consumers.

Image 17.5 The Leadership in Energy and Environmental Design certification program identifies green buildings that use less energy.

User:Drums600 / Wikimedia Commons / CC-BY-SA-3.0

Image 17.6 Changes in agriculture will need to happen on different scales, including growing more food in cities, like at this rooftop vegetable garden.

- Adopting green technologies: Whether it be light bulbs, energy-efficient appliances, or more fuel-efficient modes of transportation, the adoption of devices that consume less energy will be essential to reduce total demand (Image 17.7). However, this must not be seen as a silver bullet because researchers have found that more efficiency does not always reduce consumption; rather, it can open the door for people to adopt more energy-consuming devices.
- Lifestyle changes: Herein lies the key in that none of the above solutions will work unless we educate the public to make changes that promote a less energy-intensive lifestyle. Many of these changes are also costly, and thus it will require political will to make them available to those who cannot afford to make all these changes on their own (Image 17.8).

Solely regarding the production of energy, there are two ways to look at changes. One way is to move to more renewable energy and the other is to improve energy efficiency. Both are necessary to make the system *more* sustainable (achieving a completely sustainable system will be a long and difficult process, if it is achievable at all). As difficult as these changes may seem, taking a policy approach that favors more sustainability over time will be essential to overcome many of the negatives that encompass our current energy system.

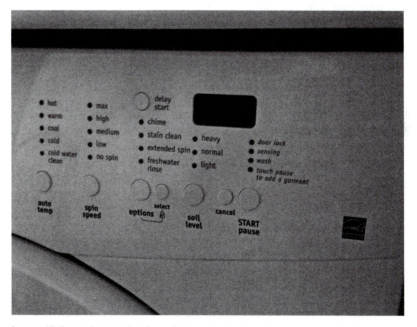

Image 17.7 Appliances that bear the Energy Star logo have been certified to be more energy efficient.

Image 17.8 Walking to work is just one lifestyle change that can impact energy sustainability.
Dragon Images / Shutterstock

Energy Switching

Achieving a stronger, more resilient, and environmentally friendly energy system is far more complex than just throwing off fossil fuels. If humankind were to stop using fossil fuels tomorrow, the global economy would collapse, and it is doubtful we could even produce enough food for the billions of people worldwide. We are nowhere near the point of developing efficiency gains that could take care of all our challenges, and even then, these changes would have to occur in each sector. Thus, it is important to think of energy switching through the lens of consuming sectors of the economy. We cannot power jet airliners or produce steel with windmills or solar panels.

When we look at the energy-switching paradigm, it is helpful to examine different resources and best understand how alternatives apply in each case. For instance, coal remains primarily of importance in developed countries for electricity production and metallurgy. Any substitute for a coal-fired power plant must provide enough power and reliability to replace it; thus, wind and solar by themselves are not enough until we develop energy storage systems capable of storing excess power for when the wind is not blowing or the sun is not shining. Nuclear power may be able to help close some of the gap, but it would require a massive expansion just to replace coal in large countries such as China, India, and the United States, an unlikely proposition because of the political controversies and high upfront costs surrounding nuclear. In metallurgy, especially steel making, coal is needed for making the steel, so until we move beyond using steel, we will always need coal. Coal or natural gas is also necessary to provide the high heat required for steel making.

Oil presents us with a different set of challenges for energy switching because its primary use is in transportation and as an industrial feedstock, rather than electricity generation (at least in developed countries). Because of its energy density as a fuel for transportation, the challenges to dramatically curtail oil consumption are far greater than that for coal. Biofuels currently only make up a minute percentage of liquid transportation fuel. Ethanol is less energy dense than oil and requires vast amounts of land, water, and fertilizer for production. Biodiesel also requires vast amounts of land, much of that coming from rainforest areas, to increase production. Even with a major breakthrough to produce cellulosic ethanol or algae ethanol, it is unlikely that we can grow enough to meet our needs. The other option is moving to battery-powered transportation. New breakthroughs may lead to the vast majority of personal cars being battery powered at some point. However, electric vehicles will have to develop longer ranges and come at a reduced cost before we see widespread adoption; it may take at least two decades for most of the current gasoline-powered cars to come off the road once such a transition occurs. For aviation, this challenge is even greater. Batteries are heavy and currently not well suited for powering a jet aircraft. Aviation will likely remain an oil user long after we have mostly abandoned it as a ground transportation fuel. And even if all transportation transitions from oil, we will still need some for the production of plastics and other crude oil–derived products. Like coal, oil is not likely to completely disappear as an energy resource.

Unpredictability as to Our Energy Future

The two examples of energy switching help illustrate the myriad questions we must consider as we discuss our energy future. New technologies will appear, geopolitical shifts could lead to massive changes in resource prices, and various political and economic choices will be made that help shape the global energy system of the future. Navigating these changes and transitioning toward a cleaner, more sustainable energy system will prove to be one of the great challenges of our time. With a growing world population, accelerating climate change, and continued geopolitical strife underpinned by energy resources, major energy shifts will be necessary. How long they take, which places lead them, and ultimately which resources and technologies win out are anybody's guess, but with the right academic foundations in place, you can be more than a passive observer as humankind moves into its next energy era.

SI and Customary Units and Their Conversions

This appendix provides a table of units and their conversion from older units to Standard International (SI) units.

Length

Metric Measure

1 kilometer (km)	=1000 meters (m)
1 meter (m)	=100 centimeters (cm)
1 centimeter (cm)	=10 millimeters (mm)

Nonmetric Measure

1 mile (mi)	=280 feet (ft)
	=1760 yards (yd)
1 yard (yd)	=3 feet (ft)
1 foot (ft)	=12 inches (in)
1 fathom (fath)	=6 feet (ft)

Conversions

1 kilometer (km)	=0.6214 mile (mi)
1 meter (m)	=3.281 feet (ft)
	=1.094 yards (yd)
1 centimeter (cm)	=0.3937 inch (in)
1 millimeter (mm)	=0.0394 inch (in)
1 mile (mi)	=1.609 kilometers (km)
1 foot (ft)	=0.3048 meter (m)
1 inch (in)	=2.54 centimeters (cm)
	=25.4 millimeters (mm)

Area

Metric Measure

1 square kilometer (km²)	=1,000,000 square meters (m²)
	=100 hectares (ha)
1 square meter (m²)	=10,000 square centimeters (cm²)
1 hectare (ha)	=10,000 square meters (m²)

Nonmetric Measure

1 square mile (mi²)	=640 acres (ac)
1 acre (ac)	=4840 square yards (yd²)
1 square foot (ft²)	=144 square inches (in²)

Conversions

1 square kilometer (km²)	=0.386 square mile (mi²)
1 hectare (ha)	=2.471 acres (ac)
1 square meter (m²)	=10.764 square feet (ft²)
	=1.196 square yards (yd²)
1 square centimeter (cm²)	=0.155 square inch (in²)
1 square mile (mi²)	=2.59 square kilometers (km²)
1 acre (ac)	=0.4047 hectare (ha)
1 square foot (ft²)	=0.0929 square meter (m²)
1 square inch (in²)	=6.4516 square centimeters (cm²)

Volume

Metric Measure

1 cubic meter (m³)	=1,000,000 cubic centimeters (cm³)
1 liter (l)	=1000 milliliters (ml)
	=0.001 cubic meter (m³)
1 milliliter (ml)	=1 cubic centimeter (cm³)

Nonmetric Measure

1 cubic foot (ft³)	=1728 cubic inches (in³)
1 cubic yard (yd³)	=27 cubic feet (ft³)

Conversions

1 cubic meter (m³)	=264.2 gallons (US) (gal)
	=35.314 cubic feet (ft³)

1 liter (l)	= 1.057 quarts (US) (qt)
	= 33.815 fl uid ounces (US) (fl oz)
1 cubic centimeter (cm³)	= 0.0610 cubic inch (in³)
1 cubic mile (mi³)	= 4.168 cubic kilometers (km³)
1 cubic foot (ft³)	= 0.0283 cubic meter (m³)
1 cubic inch (in³)	= 16.39 cubic centimeters (cm³)
1 gallon (gal)	= 3.784 liters (l)

Mass

Metric Measure

| 1000 kilograms (kg) | = 1 metric ton (t) |
| 1 kilogram (kg) | = 1000 grams (g) |

Nonmetric Measure

1 short ton (ton)	= 2000 pounds (lb)
1 long ton	= 2240 pounds (lb)
1 pound (lb)	= 16 ounces (oz)

Conversions

1 metric ton (t)	= 2205 pounds (lb)
1 kilogram (kg)	= 2.205 pounds (lb)
1 gram (g)	= 0.03527 ounce (oz)

| 1 pound (lb) | = 0.4536 kilogram (kg) |
| 1 ounce (oz) | = 28.35 grams (g) |

Pressure

| standard sea-level air pressure | = 1013.25 millibars (mb) |
| | = 14.7 lb/in² |

Temperature

To change from Fahrenheit (F) to Celsius (C)

$$°C = °F - 32/1.8$$

To change from Celsius (C) to Fahrenheit (F)

$$°F = °C - 1.8 + 32$$

Energy and Power

1 calorie (cal)	= the amount of heat that will raise the temperature of 1 g of water 1°C (1.8°F)
1 joule (J)	= 0.239 calorie (cal)
1 watt (W)	= 1 joule per second (J/s)
	= 14.34 calories per minute (cal/min)

Glossary

acid rain—Precipitation that is abnormally acidic, mostly caused by sulfur and nitrous oxides released as air pollution.

active solar heating—Solar heating, usually rooftop, of water that is used for climate control or hot-water systems on a building.

air pollution—The release of substances into the air that negatively impacts people, the built and natural environments, and/or the climate system.

albedo—The reflection of some solar radiation directly into outer space. Whiter colored surfaces produce a higher proportion of albedo reflection than darker surfaces.

algae—Single-celled organisms that use photosynthesis to produce their energy, usually living in water. Some research into biofuels look to algae for its fast-growing properties and the lack of land needed to grow it as a means to produce a plant-based biofuel.

alpha particles—A helium nucleus (two protons and two neutrons) released during radioactive decay. It is the least penetrating form of radiation.

alternating current—An electric current that changes direction periodically and is the form of electricity transmitted to most customers in the United States.

ambient air pollution—Outdoor air pollution found in areas where humans burn fuel or waste leading to the release of chemicals into the air.

anemometer—Device used to measure wind speeds.

anthracite coal—*Hard coal*, which is from 86 to 97 percent carbon. It burns slowly, carries a high price, and is used primarily in residential and industrial heating, but not for electricity generation.

Anthropocene era—New geologic era that refers to the period of time where human activity has overtaken global natural systems as the primary driver of change, often referring to the period since the Industrial Revolution.

anthropogenic—Human-caused process; often used in reference to humankind's role in contributing to climate change.

Arctic amplification—A phenomenon wherein oceanic and atmospheric circulation patterns cause climatic changes to be exaggerated in the Arctic.

associated gas—Natural gas products found suspended in crude oil, which are removed at the well and flared, captured, or reinjected into the oil reservoir.

barrel—Measurement used for oil measuring 42 U.S. gallons.

barrel of oil equivalent—The amount of energy in one barrel of oil, approximately 5.8 million British thermal units.

beta particles—An electron or positron emitted during radioactive decay.

binary geothermal power plant—A geothermal plant that uses a low-boiling-point liquid adjacent to hot water to create steam for the turbines.

bioaccumulation—The concentrating of toxins in the body of higher order (i.e., predatory) animals, such as the concentration of mercury in large fish. Many of these bioaccumulates enter the water system via air and water pollution related to the burning of coal or heavy fuel oils.

biodiesel—A biologically sourced fat that can be used in lieu of conventional diesel fuel in a diesel engine.

biofuel—Any fuel that is derived from organic materials (usually plant or animal), often used to refer specifically to organically derived liquid fuels used to replace fossil fuels for transportation, although wood and other biomass burned directly is also a biofuel.

biogas—*See* biogenic methane

biogenic methane—Methane formed during the decomposition of organic materials and released by bacteria.

bitumen—A black and glossy soft solid or highly viscous liquid that is a form of crude oil, also known conventionally as asphalt.

bituminous coal—A hard coal composed of 47–86 percent carbon used in electric generation and industrial processes.

black lung disease—Condition caused by long-term exposure to coal dust, especially in coal mines, also known as pneumoconiosis. The introduction of respirators has cut back, but not eliminated, black lung disease in more advanced coal mines.

breeder reactors—Nuclear reactors that use more highly enriched uranium and produce higher amounts of plutonium, also known as fast neutron reactors.

British thermal unit—The amount of energy needed to raise one pound of water by one degree Fahrenheit.

butane—The second heaviest carbon compound commonly found in natural gas, often used in refining, as a chemical feedstock, or as a light liquid fuel (i.e., in cigarette lighters).

CANDU reactor—Canadian deuterium uranium reactor, which uses unenriched uranium to produce a fission nuclear reaction for a power plant.

carbon capture and sequestration—Process of capturing carbon dioxide in the exhaust of a fossil fuel–burning plant and sequestering the carbon dioxide deep underground rather than releasing it into the atmosphere.

carbon dioxide (CO$_2$)—An odorless and colorless nontoxic gas released during the combustion of carbon-based fuels that is the most common greenhouse gas and one of the main ingredients in fossil fuel exhaust.

carbon monoxide—An odorless and colorless toxic gas formed during incomplete combustion of carbon-based fuels.

carbon neutral—Refers to any process wherein the net effect on atmospheric carbon dioxide is zero because of the burning of carbon-based fuels that incorporate carbon recently fixed from the atmosphere, rather than fossilized and stored in the earth's surface.

carboniferous—Geologic era meaning coal bearing, dating from 359–299 million years ago when the earth had a much higher concentration of carbon dioxide in the atmosphere, when much of the planet's coal was formed.

cartel—An agreement between competing firms or countries to engage in price setting, as attempted by the Organization of the Petroleum Exporting Countries to set global oil prices.

catalytic converter—Motor vehicle part in the exhaust system that treats exhaust by passing it through a filter where aluminum oxide, rhodium, palladium, and platinum serve as catalysts to chemical reactions in which harmful gases are converted into carbon dioxide, water vapor, and nitrogen.

cellulosic ethanol—Ethanol fuel produced using the starchy and woody parts of a plant such as stalks, rather than using seeds and other sugary and oily parts of the plant.

chain reaction—Nuclear reaction that sustains itself via thrown-off neutrons impacting additional atomic nuclei.

chemical energy—Energy derived through the breaking or forming of chemical bonds, such as through the combustion of gasoline or the process undergone in a battery.

chlorofluorocarbons (CFCs)—Complex carbon-chain gases that are a major air pollutant, especially correlated with the destruction of the ozone layer.

circuit—A closed system that allows for the flow of electrical current.

Clean Air Act—A 1963 U.S. law that limited the amounts of several air pollutants that could be emitted and gave the federal government jurisdiction over regulating air pollution emissions.

closed fuel cycle—Nuclear fuel cycle in which spent nuclear fuel is reprocessed and unused U-235 is reprocessed and used again in nuclear fuel.

coal—Fossil fuel composed of carbon and other elements, which is formed from decomposed plant materials that are trapped under the earth and have undergone intense heat and pressure for thousands to millions of years.

coal-bed methane—Methane deposits associated with coal seams that can be separately captured via drilling and used like any other natural gas.

Comprehensive Nuclear Test Ban Treaty—A 1996 treaty that bans the testing of nuclear weapons.

compressed natural gas (CNG)—Methane and other natural gases that are compressed into a storage tank, often for small-scale use such as a fuel for motor vehicles.

concentrated solar power (CSP)—The harnessing of solar power to produce electricity via thermal generation.

concentrating photovoltaics (CPV)—Solar photovoltaic that uses mirrors or lenses to concentrate the sun's rays onto a smaller surface and work more efficiently than normal photovoltaics, but at higher cost.

continental crust—The earth's crust on plates that are primarily land and the continental shelf, containing most fossil fuel deposits.

control rods—Rods in nuclear reactors that control the rate of the nuclear chain reaction, often composed of boron, silver, cadmium, or indium.

corn ethanol—Biofuel derived from the fermentation of corn kernels and their distillation into grain alcohol.

criteria pollutants—Six highly dangerous air pollutants regulated in the United States by the Environmental Protection Agency: sulfur dioxide, nitrous oxides, ozone, carbon monoxide, particulates, and atmospheric lead.

crude oil—Viscous liquid fossil fuel that is the unrefined form of petroleum (oil) and contains a mix of various hydrocarbons from natural gas to tar in different amounts depending on source location.

crust (earth)—The outermost solid layer of the earth, consisting of various rocks.

dam—An obstruction along a flowing body of water that holds water in a reservoir.

Darrieus turbines—A type of vertical-axis wind turbine that uses winglike blades to reduce stress on the frame.

dead pool—Condition wherein a reservoir level falls below the lowest intake valves on a dam, leading to no water flow through the dam downstream.

deuterium—Isotope of elemental hydrogen that contains one proton and one neutron, thus having the atomic weight of 2, whereas most hydrogen has no neutron and thus an atomic weight of 1.

diesel engine—Internal combustion engine that achieves ignition from high pressures and temperatures, often used in trucks and larger vehicles such as ships and trains. It utilizes a different refined product than do standard gasoline engines.

diesel fuel—Refined petroleum fuel or biofuel used in diesel engines.

diesel vehicles—Motor vehicles that utilize a diesel engine instead of a standard internal combustion engine or electric motor.

dry steam plant—Geothermal power plant that utilizes naturally occurring steam to turn turbines, rather than injecting water for such use.

Dutch disease—Economic condition in which the growth of one sector inhibits or damages other sectors, often seen in petro-states. The term was first coined in reference to economic conditions in the Netherlands after the discovery of natural gas in the 1970s in the North Sea.

EIA—*See* U.S. Energy Information Administration

electrical energy—Energy in the form of electrons that flow via a current.

electrical generator—Electromagnetic device that uses the rotation of magnets to create an electrical current.

electrical current—The flow of electrons through a circuit.

electricity—Phenomenon that comes from the flow of an electrical charge, which when flowed through a conducive material can be consumed as energy for human use.

electrolysis—Process of separating hydrogen and oxygen by breaking up water with the use of an electric current, used primarily to collect the hydrogen.

energy—The ability to do work. For the purposes of studying human consumption of energy, this refers primarily to heat energy and kinetic energy.

energy crops—Plants grown for use as biofuels including sugarcane, corn for ethanol, and some oily seed crops for biodiesel.

Energy Information Administration—*See* U.S. Energy Information Administration

energy switching—Transitioning from the use of one energy source to another, currently associated primarily with the transition from dirty fossil fuels to cleaner energy sources, including alternative energy resources.

Environmental Protection Agency (EPA)—Agency of the U.S. government charged with enforcing environmental laws and regulations including the Clean Air Act and Clean Water Act.

ethane—The second lightest natural gas, used primarily as a chemical feedstock.

ethanol—Also called ethyl alcohol. This colorless liquid is produced as a biofuel substitute for gasoline.

evacuated tube collector—Solar hot water system popular in colder climates, in which vacuum-sealed glass tubes conduct water through a rooftop solar water-heating system.

feed-in tariffs—Guaranteed minimum price for certain forms of electricity generation to ensure that renewables such as wind and solar net a certain price when fed into the electrical grid.

fish ladders—Series of steps on a dam with water flowing down them that allow for migratory fish to jump up the ladder to travel upstream, commonly found in salmon habitats.

fixing (carbon)—The process by which plants and other photosynthetic life forms remove carbon dioxide from the air and "fix" it into another carbon-based molecule.

flash steam plant—The most common type of geothermal power plant, which injects water into a geothermal reservoir wherein it flashes to steam point and is piped back to the surface and through a turbine.

flat plate collector—Active solar water heater that uses a sheet of glass on the front, an air space, and then a flat, dark-colored solar absorber with tubes running behind it to absorb the heat and then pipes the hot water or other heated substance into the building for use or storage in a tank.

fly ash—A residue of fine particulate matter released in flue gas from the burning of coal.

fracking—*See* hydraulic fracturing

fracking fluid—Mixture of mostly water with detergents and proppant (sand or other similar material) injected into a well as part of the process of hydraulic fracturing.

fractional distillation—Step in the process of oil refining wherein crude oil is heated and settles into various compounds by weight and can be sorted.

fossil fuels—Energy resources that are derived from organic materials of the distant past that are trapped and transformed within the earth's crust, consisting of coal, oil, and natural gas.

fuel cell—Device that produces electricity by passing hydrogen and oxygen through an anode and cathode to produce electricity while producing water vapor as the sole waste product.

gamma particles—The hardest to detect and most penetrating form of radiation, consisting of a high-energy photon that can only be blocked by dense substances such as lead or thick concrete.

gasoline—Refined petroleum product primarily used for motor fuel in cars, trucks, and other small machines.

geopolitics—Study of the political relationships of power between countries and other significant actors on the global stage. Relevant to energy resources in the study of global competition over energy resources, especially oil, natural gas, nuclear fuel, and hydrologic resources that cross international borders.

geothermal energy—The energy drawn from the intense heat created by the interior of the earth. One of only three energy resources not originally drawn from the power of the sun.

geothermal gradient—The change in temperature correlated with depth at any given point on the earth's crust.

geothermal reservoir—Location where there is greater heat under the ground that can be used for more intensive forms of geothermal power.

graphite—A crystalline form of carbon used in the manufacture of nuclear fuel pellets.

green power—Colloquial term that refers to forms of power generation that have a lower impact on the environment, most often associated with wind and solar, along with other renewable power sources.

greenhouse effect—Process by which some solar heating is trapped by greenhouse gases in the atmosphere, rather than radiating into outer space.

greenhouse gas (GHG)—Gases in the atmosphere that trap solar radiation, including carbon dioxide (the primary greenhouse gas), methane, nitrous oxides, and chlorofluorocarbons.

grid storage—System wherein electricity that has already been generated is stored within a power grid, such as in a pumped storage facility.

ground source heat pump—*See* heat pump

gushers—Oil wells that in the early era of oil exploration would hit an oil reservoir, upon which oil would gush from the top of the rig until capped.

half-life—The amount of time it takes for half an amount of an unstable element to break down into two lighter elements, which can vary from nanoseconds for some of the highest elements on the periodic table to millions of years.

heat energy—Energy released in the form of heat.

heat pump—Geothermal technology that takes advantage of the stable temperatures encountered beneath the surface, including in areas with a low geothermal gradient, to use for heat exchange for both building heating and cooling.

heating oil—Heavy petroleum product used in some places to provide residential or commercial building heating.

heavy fuel oil—One of the heaviest substances other than tar in crude oil, which is used primarily for electricity production and overseas shipping, also known as bunker fuel.

heavy-water reactor—Nuclear reactor that uses heavy water, which contains more deuterium in it, as the moderator during a nuclear chain reaction in the reactor.

high-level waste—The most radioactive parts of radioactive waste, consisting mostly of spent nuclear fuel and various other chemicals removed during fuel processing.

highly enriched uranium—Uranium that has been processed so that at least 20 percent is U-235 and, for weapons, up to 90 percent U-235.

hot spots—Locations in the earth's crust where the crust is thin and a magma chamber exists closer to the surface, such as the area near and directly under Yellowstone National Park.

Hubbert curve—Bell curve showing the production of oil at a given well or field or extrapolated out to an entire country or the world, which forms the basis for the peak oil theory.

Hubbert's peak—*See* Hubbert curve

hybrid car—Motor vehicle that utilizes both a conventional gasoline internal combustion engine and an electrical motor to store excess energy and reduce gasoline consumption.

hydraulic fracturing—Process wherein a horizontal well is drilled into an oil- or gas-bearing rock layer and fracking fluid is injected at high pressure to break apart rock and allow for the flow of oil and/or gas to the wellhead, allowing the production of both oil and gas in locations where either a limited amount or no fuel could be produced otherwise; also known as fracking.

hydrocarbon runoff—The release, usually by leakage, of hydrocarbons into the environment during any phase from the well to final consumption of the resource.

hydrocarbons—Organic molecules consisting primarily of carbon chains with hydrogen attached throughout that form the basis of crude oil and natural gas.

hydroelectricity—The production of electricity by turning a turbine with flowing water.

hydrogen—The lightest of all elements that can be used as an energy resource either via fuel cells or by direct combustion, either of which releases only water vapor at the end.

hydropower—The deriving of any form of energy using flowing water, including not only electrical generation but also the direct use of kinetic energy for various processes such as pumping water or grinding grain.

IEA—*See* International Energy Agency

indoor air pollution—The release of air pollutants indoors, which in the energy context primarily is caused by wood- or coal-burning stoves inside homes.

induced seismicity—Manmade earthquakes caused by the injection of fluids into the ground, commonly associated today with wastewater injection at oil and natural gas fields, although carbon sequestration could also cause this in future.

industrial revolution—The transition in the economy from most goods being produced by hand to being produced by machines.

industrial smog—Smog caused by the burning of coal or other fossil fuels for electrical generation and industrial production.

inertial confinement—Fusion caused by the collapse of light elements as being tested at the National Ignition Facility.

inner core—Innermost layer of the earth, consisting primarily of superheated nickel and iron.

in situ production—The production of something in place; in energy usually referring to the injection of steam into tar or oil sands underground, liquefying them, and bringing them to the surface. This term could also be applied to oil shales.

Intergovernmental Panel on Climate Change (IPCC)—International organization that brings together scientists studying the climate and has led to the adoption of various climate-related treaties and agreements.

intermediate wastes—Nuclear waste that consists of any substance that came into contact with the nuclear core, such as the fuel assembly.

internal combustion engine—Any engine that burns refined petroleum or natural gas products, such as the engines in automobiles.

International Atomic Energy Agency (IAEA)—International organization based in Vienna that regulates worldwide use of nuclear technology for peaceful purposes and enforces the Nuclear Non-proliferation Treaty.

International Energy Agency (IEA)—Paris-based international organization that coordinates on international energy issues between countries in the Organisation for Economic Co-operation and Development (developed countries) and maintains international energy statistics.

jet engine—Internal combustion engine that creates a jet or thrust to provide propulsion; can refer both to turbofan engines used by many aircraft and to rocket engines.

jet fuel—Fuel blend that is commonly used for aircraft engines and is slightly denser than gasoline.

joule—Unit of energy that describes the amount needed to move one newton one meter.

kerogen—The source substance for petroleum, which is found in sedimentary rocks and is composed of various organic compounds that forms into crude oil and natural gas when placed under sufficient heat and pressure for thousands to millions of years.

kinetic energy—The energy of motion.

law of conservation of energy—Law of physics that describes that a total amount of energy in a fixed system remains constant over a given time, and can only change state rather than be destroyed.

lead—A soft, heavy, metallic element that is dense and is a neurotoxin. In energy resources, lead is commonly used to contain radiation in nuclear contexts and in the past was used as a gasoline additive, which led to the widespread release of lead into the atmosphere. It continues to be released with the burning of coal.

LEED certification—Leadership in Energy and Environmental Design designation given to buildings that meet requirements for energy and water efficiency in both the construction and the operation of the building.

light-water reactor—Nuclear reactor that uses normal fresh water as the coolant and neutron moderator. This is the most commonly used form of nuclear reactor.

lignite—"Brown coal" consisting of 25–35 percent carbon, which is a dirty-burning fuel with high water content often used for electricity generation.

liquefied natural gas (LNG)—Methane that has been supercooled and compressed into a liquid for overseas transport, which consumes a significant amount of energy both to liquefy and again to regassify once transit is completed.

liquefied petroleum gases—Lightweight fuels such as propane, ethane, and butane, which can be easily stored in liquid form and used for purposes such as cooking gas.

longwall mining—Coal-mining technique wherein one face of the coal seam is removed at a time with a conveyor system in place along the mining face, with areas left intact to support the rock layers above, or areas are allowed to collapse once the coal is removed.

low-level waste—Most common form of radioactive waste (over 90 percent) that is composed of any materials that were subjected to radiation either in a nuclear plant or during medical use.

Manhattan Project—Top-secret U.S. government project during World War II that developed the first nuclear weapons, including those used on Japan during the war.

mantle—The layer of the earth located beneath the crust and above the core that is hundreds of miles thick and consists of silicate minerals that are superheated and exist in a plastic form.

mercury—Liquid metal element that is a neurotoxin and is often released during coal combustion; may also be found in some crude oils in small amounts.

methane—Organic combustible compound also commonly called natural gas that is the lightest hydrocarbon and the cleanest burning, in that it only releases carbon dioxide and water vapor when combusted.

methane hydrates—Methane molecules trapped inside water crystalline structures both along the margins of the continental shelf and in some permafrost soils. There is more methane in these hydrates than in all natural gas deposits in rock worldwide.

mountaintop removal—Coal-mining technique wherein the top layers of a mountain with a coal seam are detonated and removed to the valley below to access the coal, which leads to dramatic and irreversible changes in a region's topography.

National Ignition Facility (NIF)—Facility in California that is part of the Lawrence Livermore National Laboratory and has built the world's largest laser, in part for fusion experiments.

natural gas—Fossil fuel energy resource composed primarily of methane for a variety of uses, including electrical generation, home heating, cooking, and transportation.

nitrous oxides (NO_x)—Gases produced in combustion, primarily in motor vehicles, which lead to acid rain and smog production.

nonrenewable energy resources—Generic term for any energy resource that will not regenerate within a short period of time and includes all fossil fuels and nuclear energy.

Not in My Backyard—Sentiment that opposes the location of undesirable facilities, such as power plants, near residents.

nuclear deterrence—Geopolitical theory that when multiple countries have nuclear weapons, they will not use them against others for fear that they will also be targeted with nuclear weapons.

nuclear fusion—Process wherein smaller elements fuse together to form larger elements, with the fusing of hydrogen to form helium being the most common. Although never achieved in a sustainable way by humans, nuclear fusion within stars is the most common form of energy production in the universe.

Nuclear Non-proliferation Treaty (NPT)—Treaty signed in 1968 that called for no additional countries to develop nuclear weapons beyond those that had already tested such weapons—the United States, the Soviet Union (Russia inherited their nuclear rights), the United Kingdom, France, and China—and created an international framework for the peaceful use of nuclear power.

nuclear power—The production of electricity via the fission of heavy elements, primarily uranium.

oceanic crust—Thinner and often younger crust of the earth found on oceanic plates, compared to the continental crust found on the remaining plates.

octane rating—Standard measure of the amount of pressure a fuel can withstand before igniting, with higher performance gasoline engines needing a higher rating, whereas diesel has lower octane ratings than conventional gasoline engines.

OECD—*See* Organisation for Economic Co-operation and Development

offshore drilling—Oil- or gas-drilling operations that are conducted over water on a platform that can either be affixed to the ocean floor or float, depending on the depth of the water.

oil—Fossil fuel formed from phytoplankton that is liquid to semisolid at room temperature; the world's most consumed energy resource.

oil refinery—Industrial facility that breaks down (or cracks) crude oil into its various component compounds for consumption as a fuel or for use as a feedstock for other purposes.

oil shale—Oil-bearing rock type that is denser than sandstone and requires heating or fracking to produce the oil within.

OPEC—*See* Organization of the Petroleum Exporting Countries

open fuel cycle—Nuclear fuel cycle wherein uranium is only used once and then not reprocessed for continued use. The United States utilizes the open fuel cycle.

Organisation for Economic Co-operation and Development (OECD)—Paris-based international organization of the industrialized and developed economies, whose membership is often considered synonymous with the developed or first world.

Organization of the Petroleum Exporting Countries (OPEC)—Vienna-based international organization or cartel that consists of many of the world's major oil exporters and tries to coordinate on overall oil production and influence the price of oil globally.

outer core—Second layer of the earth from the center that is liquid iron and nickel and is about 2,300 kilometers (1,400 miles) thick.

overtopping—Dangerous condition on a dam wherein the level of the reservoir rises higher than the dam and the reservoir starts to flow uncontrolled over the top of the dam. This can lead to a catastrophic dam failure.

ozone—Gaseous compound consisting of three oxygen molecules (most oxygen consists of two) that in the stratosphere helps protect against ultraviolet radiation from the sun, but at surface levels is damaging to human health.

Pacific Ring of Fire—Large concentration of volcanoes found around the edges of the Pacific Ocean, resulting from the subduction of the Pacific Plate under numerous surrounding plates. Many of the best places for geothermal development lie along the Ring of Fire.

palm oil—Plant-based oil produced from the oil palm seed that is often used as a biodiesel, along with various other nonenergy purposes. Its production is one of the primary drivers of tropical rainforest destruction, especially in Southeast Asia.

parabolic trough—Solar thermal plant that uses parabola-shaped mirrors to direct sunlight on tubes containing a superheated liquid that can be used to boil water and produce steam to generate electricity in the conventional manner.

Partial Test Ban Treaty—A 1963 nuclear treaty that limited states' ability to test nuclear weapons to underground testing, whereas many early tests were conducted above ground.

particulates—Air pollutant consisting of microscopic solids that embed themselves in the lungs and create dust and grime downwind of their creation, often from unscrubbed coal chimneys and wood burning.

passive solar heating—Simplest form of solar energy use wherein windows and substances with a high specific heat are used to help warm an indoor space.

peak oil theory—Theory first postulated by the geologist M. King Hubbert that assumes that all oil production occurs on a bell curve for wells, fields, and entire countries, eventually leading to a global peak and then decline in oil production that is irreversible.

peat—Organic material composed of dead plants that become trapped in a low-oxygen and water-rich environment, which serves as the base material to form coal. Peat can also be dried and burned as an energy resource, but it is of poor quality.

penstock—Openings in a dam that can allow water to flow through the dam.

pentane—The heaviest compound in crude oil to exist as a gas at temperatures near room temperature, often used as a solvent.

percussion drilling—Form of oil and gas drilling used in earlier periods wherein a drill would be raised and struck repeatedly downward; this technique has been replaced by rotary drills.

petroleum—Generic term for crude oil and its derivatives.

photobioreactor—System for growing algae in a clear material where carbon dioxide gas is pumped into water and algae.

photochemical smog—Air pollution caused by the reaction of sunlight with volatile organic compounds and/or nitrogen oxides.

photovoltaic effect—Effect wherein sunlight is directly converted into an electrical current in a solar cell.

photovoltaics (PV)—Generic term for any solar cells used in the production of electricity.

phytoplankton—Microorganisms found in water that use photosynthesis to produce their energy and serve as the organic feedstock for crude oil and gas formation.

pillar mining—Coal-mining technique wherein columns of coal are left intact within a mine to keep the mine from collapsing; this technique has been mostly replaced by longwall mining in developed countries.

pipeline—Pipes either on the surface or buried underground that are used to transport liquids and gases including crude oil and natural gas, as well as many other substances outside the energy realm.

pitchblende—Radioactive mineral also known as uraninite that is one of the primary minerals mined for the production of uranium.

plasma—The fourth physical state of matter (other than solid, liquid, and gas) wherein molecular bonds break down, as found in stars and other fusion reactions.

plug-in hybrid car—Automobile with both a gas and an electric motor where the batteries for the electric motor can be charged off the power grid by being plugged in when the vehicle is parked.

potential energy—The energy possessed within an object due to position, chemical makeup, or other factors.

Powder River Basin—Geological region of Wyoming that contains the most productive coal mines in the United States.

power—In physics, the rate of doing work measured in watts.

producing formation—Sedimentary rock formation that has trapped oil or natural gas that is being produced through wells.

propane—The middle-weight natural gas that is often used for cooking gas, among other uses.

proppant—Solid substance, often sand, that is added to hydraulic fracturing fluid to keep the cracks formed in rock open to thereby allow the flow of oil or natural gas to the wellhead.

pumped-storage hydroelectric—Energy storage system where excess energy, usually electricity, is utilized to pump water up a hillside and can then be allowed to flow down the hill and through a hydroelectric turbine when energy demand is higher.

radiant energy—The energy in solar radiation such as that used in solar power.

radiation—The emission or transmission of energy in particles or waves.

radiative forcing—The difference between solar energy absorbed and radiated back into outer space.

range anxiety—Concerns over the limits in distance a vehicle can travel before running out of energy, which is of concern for electric vehicles because of the inability to quickly refuel.

renewable energy resources—Generic term for any energy resource that is renewed by natural systems in a short amount of time, which applies to all forms of energy consumed by humans except fossil fuels and nuclear energy.

reprocessing—The treatment of spent nuclear fuel to remove remaining U-235 and plutonium, which can be made again into nuclear fuel.

reservoir (oil)—Concentration of oil deposit underground.

reservoir (water)—Manmade lake behind a dam along a river.

reservoir pressure—The amount of pressure occurring within a petroleum reservoir that aids in the flow of oil and gas to the surface and, once a certain amount is relived, causes engineers to need to pump or flush additional oil and gas to the surface.

rotary drilling—Oil- and gas-drilling technique that uses rotary drills to bore a well into the ground to a reservoir of oil and/or gas.

run-of-the-river system—Hydroelectric system placed directly into flowing water rather than utilizing a dam, usually done on a smaller scale than hydroelectric dams.

salinization—Process wherein salts are leached to the surface of soils primarily in arid soil types, found downstream of dams that cause regular flooding to subside along a river.

sandstone—Sedimentary rock that consists primarily of sand and therefore is highly porous and often contains water or petroleum.

scrubbers—System where flue gases can be passed through to remove harmful compounds including sulfur dioxide, nitrous oxides, heavy metals, and particulates before the remaining gases are sent up a chimney into the atmosphere.

sedimentary rock—Class of rocks formed by sediments that collect and are buried and turned to rock. All fossil fuels are formed in sedimentary rocks.

seismic imaging—Imaging of subsurface rock layers using sound waves that bounce back at different frequencies depending on the type of rock they hit.

shale—Sedimentary rock, consisting primarily of clays and silts, which has a very low porosity.

siltification—The deposition of silts in a reservoir caused by the slowing or stopping of water flow, which eventually result in the reservoir filling with silt.

slurry—Mixture of a ground-up solid with water to move through a pipeline or factory, which is used to send coal via pipelines.

smog—Air pollution caused by motor vehicles and industrial processes, including coal- and oil-fired power plants, that leads to severe degradation of local ground-level air and a chemical fog.

solar energy—Any form of energy that directly uses the heat of solar rays or the photovoltaic effect.

solar gain—The amount of solar energy absorbed as heat in active and passive solar heating systems.

solar power tower—Solar power plant that uses a vast array of mirrors aimed at the top of a tower where molten salt is heated and then used to produce steam and electricity via the conventional method.

static electricity—An imbalance of electric charges between objects.

steam—Gaseous water produced by boiling water, which is then used to drive turbines in conventional thermal power plants.

steam engines—An engine that uses steam to drive a piston. Many early engines of the industrial revolution were steam engines, as were the engines that powered early railroads and many large ships that used coal as their power source.

steam reforming—Method of producing hydrogen from natural gas by mixing methane with steam to produce hydrogen gas and carbon monoxide.

Stirling dish—Concentrated-heat solar system that uses a mirrored dish with a Stirling engine receiving the solar heat where it drives a piston to produce power.

stratigraphic traps—Oil traps that are created by changes in the characteristics of rock formations rather than by changes in the shape of rock formations.

stratum—Layer in a rock formation with a uniform identifiable makeup.

strip mining—The removal of a resource by digging into rock from the surface, which now accounts for a large amount of global coal production and small amounts of oil shale production.

structural traps—Oil traps that are created when a rock formation is warped into a shape that prevents oil from migrating further, such as an anticline or a salt dome.

subatomic particles—Protons, electrons, and neutrons, the component parts of an atom.

subbituminous coal—A soft, black coal that ranges from about 35 to 45 percent carbon content, is lower in sulfur, and is commonly used in electric generation.

sugarcane ethanol—Ethanol biofuel produced from sugarcane that serves as a major motor vehicle fuel in Brazil.

sulfur dioxide—Gaseous air pollutant that is the primary cause of acid rain.

surface mining—The surface removal of a substance such as coal or oil sand.

surfactant—Detergent substances that are used to reduce friction in fluids such as hydraulic fracturing fluid.

sustainability—Refers to something or some process being able to be done in a manner that will allow for its continuance in future generations.

tailings (mining)—Leftover rock removed from a mine that often contains poisonous substances that can leach into the water system.

tar sands—Sand or loose sandstone layers near the surface that contain large quantities of bitumen.

thermal reactors—Standard form of nuclear reactor that uses a slower chain reaction than breeder reactors.

thermal runoff—Hot water that flows from a thermal power plant into a neighboring body of water, thereby locally raising water temperatures.

thermogenic methane—Methane created by the breakdown of organic materials under intense heat and pressure over long periods of time, which is the primary form of methane harvested as a fossil fuel.

thermohaline circulation—The worldwide circulation of ocean currents that is currently undergoing changes from the massive melting of ice caps as part of climate change.

thermonuclear—Large-scale nuclear weapons such as hydrogen bombs.

thin-film photovoltaics (TFPV)—Photovoltaic cells that are applied as a film onto other substances such as windows or roof tiles.

tidal barrage—Tidal power system wherein a dam is built at an inlet so that water flows over it at high tide and back out through penstocks and turbines as the tide goes back out.

tidal fence—Tidal power system wherein tidal flow is forced through narrow slits, which then turn blades in an area that experiences large differences between high and low tide.

tidal turbine—Tidal power system that works like a wind turbine, with tidal flow turning an underwater propeller.

tokamak—Plasma containment system that uses a torus-shaped system with electromagnets to create a flow of plasma, which hypothetically could be used for a fusion power system.

ton of coal equivalent—Measurement of the amount of energy from burning one ton of coal. Because various grades of coal release different amounts of energy, there is no international standard.

ton of oil equivalent—Measurement of the amount of energy from burning one ton of oil, approximately 42 gigajoules.

tragedy of the commons—Economic theory wherein individuals make decisions that may be beneficial at the time, but are damaging to the overall common good.

transesterification—Chemical process used to turn plant- or animal-based fats into biodiesel fuel.

transmission loss—Loss of electrical current as electricity travels along a power line.

traps (oil)—Locations in the earth's crust where crude oil becomes trapped in a permeable layer below an impermeable layer that can then be tapped.

tritium—The heaviest and rarest form of elemental hydrogen, which contains one proton and two neutrons in the nucleus, thus being of the atomic weight 3.

unassociated natural gas—Natural-gas deposits that are found without crude oil in the same deposit.

unconventional petroleum—Any form of oil or natural gas other than simple fields such as shale or oil sands.

underground mining—Mining process where a tunnel is bored into the surface and then a substance, such as coal, is removed and brought to the surface.

uraninite—*See* pitchblende

U.S. Energy Information Administration (EIA)—Statistical branch of the U.S. Department of Energy.

volatile organic compounds—Series of organic compounds, mostly gaseous, which can contribute to smog and other forms of air pollution.

voluntary pricing schemes—An agreement on the part of corporations to put a tax-like cost on emissions of pollutants or greenhouse gases, which aims to use market forces to encourage more sustainable corporate behaviors.

waste-to-energy—System where waste products such as municipal trash are burned in a conventional thermal power plant to produce electricity.

water pollution—The introduction of dangerous substances into the water, commonly caused in energy by mining and acid rain, among other ways.

water pumps—A pump that uses energy to pressurize and move water, which in the energy realm was one of the simplest uses for windmills.

watt—Measure of the rate of energy transfer. One watt equals one joule per second.

wind—The movement and flow of gases in the atmosphere from areas of higher to lower atmospheric pressure, often driven by solar energy.

wind farm—Collection of wind turbines in one location to produce large-scale wind power.

wind power generation—The creation of electricity using wind to turn rotor blades and thereby turn a crankshaft and produce power.

wind turbine—A wind machine that captures wind to turn a turbine and produce electricity.

windmill—A wind machine that captures wind power to turn machines such as water pumps or to grind grain, among other applications.

yellowcake—Uranium ore that has been leached and milled into a yellow powder that can then be used for processing nuclear fuel.

Index